A REVISION OF THE GENUS *ASTRAGALUS* L. (LEGUMINOSAE-PAPILIONOIDEAE) IN INDIA

THE AUTHOR

Dr. Lal Babu Chaudhary, born at Sitamarhi district in Bihar in 1962, obtained his Master degree in Botany from University of Bihar, Muzaffarpur. He began his research career at Botanical Survey of India, Kolkata as a research scholar and obtained his Ph. D. degree from University of Kalyani, Kalyani, W. Bengal. Dr. Chaudhary is actively engaged in his research activities on various aspects of taxonomy and biodiversity and has completed more than a dozen research projects funded by different funding agencies. His main interest in taxonomy lies in revisionary studies of legumes apart from floristic and diversity studies. Other than *Astragalus*, he has revised many legume genera like *Argyrolobium, Indopiptadenia, Lotus, Medicago, Melilotus, Oxytropis, Parochetus, Piptanthus, Thermopsis, Trifolium, Trigonella* etc. Apart from legumes, Dr. Chaudhary has also worked on *Aconitum* (Rananculaceae), *Ficus* (Moraceae), *Phyllanthus* (Phyllanthaceae), *Podophyllum* (Podophyllaceae) and *Tricholepis* (Asteraceae). So far he has published two books and about 100 research papers in different journals of national and international repute which include more than two dozen papers alone on *Astragalus*. His recent work includes a very illustrated book 'Trees of Uttar Pradesh - Part 1'. He has also supplied materials on legumes to BSI, Kolkata for publication in 'Flora of India-legume volume' which is awaiting publication.

After completing his Ph.D., he joined CSIR-National Botanical Research Institute, Lucknow as a permanent faculty wherein he presently serves as Senior Principal Scientist. During his 30 years of research experience, he has extensively travelled throughout the country, especially in the Himalaya for plant collection and diversity studies. He is actively engaged in teaching of Ph. D. students in AcSIR at CSIR-National Botanical Research Institute, Lucknow and also guiding many research scholars for their Ph. D. work. He is a member of many scientific societies and editorial boards. He has served as secretary of 'Association for Plant Taxonomy' in past. Currently, he is working as Managing Editor of the journal 'Ethnobotany' since last four years.

A REVISION OF THE GENUS *ASTRAGALUS* L. (LEGUMINOSAE-PAPILIONOIDEAE) IN INDIA

– *Author* –

L. B. Chaudhary

Plant Diversity, Systematics and Herbarium Division
CSIR-National Botanical Research Institute
Rana Pratap Marg, Lucknow –226 001
Uttar Pradesh, India

2022

Associated Publishing Company
A Division of
Astral International Pvt. Ltd.
New Delhi – 110 002

Front Cover Photo: *Astragalus oplites* Benth. ex Parker (Photo by M. Sanjappa)

ISBN: 9789354615351

Published by : **Associated Publishing Company**®
A Division of
Astral International Pvt. Ltd.
– ISO 9001:2015 Certified Company –
4736/23, Ansari Road, Darya Ganj,
New Delhi-110 002
Ph. 011-43549197, 8130496929
E-mail: info@astralint.com
Website: www.astralint.com

Dedicated to my Teacher
Dr. M. Sanjappa
Ex Director
Botanical Survey of India, Kolkata

वै.औ.अ.प.-राष्ट्रीय वनस्पति अनुसंधान संस्थान
CSIR-National Botanical Research Institute
राणा प्रताप मार्ग, लखनऊ – 226 001, उ.प्र., भारत
Rana Pratap Marg, Lucknow - 226 001, U.P., India

प्रोफेसर एस. के. बारिक
निदेशक
Professor S. K. Barik
Director

Foreword

Critical taxonomic study is the core of biodiversity assessment. Taxonomic revisionary study is one of the strongest ways for real assessment of biodiversity or organism in question. Even the effects of climate change cannot be assessed precisely unless the majority of the species either at global level or regional level are studied elaborately.

Astragalus with about 2900 species in the world is the largest genus among angiosperms. Due to its large size and enormous morphological plasticity, the taxonomy of the genus is very complex and phylogeny is still unresolved. Although several works on the genus have been carried out at national and international levels, considering the size of the genus, these seem to be inadequate. The Indian *Astragalus* which is mainly found in the cold deserts of the Himalayas also needed a comprehensive systematic study as the genus has not been studied after the late 19th century. In this context, the present work 'A revision of the genus *Astragalus* L. (Leguminosae-Papilionoideae) in India' is a welcome step by the author.

The author Dr. L.B. Chaudhary has provided an exhaustive taxonomic account of 79 species of *Astragalus* along with the four species of the genus *Phyllolobium* which has been recently segregated from *Astragalus*. The inclusion of *Phyllolobium* in the present work is quite useful to understand the relationship between the two genera. The originality of the work based on extensive field surveys and examination of large number of previous collections housed at different herbaria is appreciable. The taxonomic notes provided for species on variations, relationship and distribution should be extremely useful to the relevant researchers. The author has also provided first hand information on phenology and habitats of the species. Since *Astragalus* is considered as an indicator species, these information may be utilized in climate change studies in the Himalayas. A large number of photographs included in the book further enhances the quality of the work and makes the identification easy.

Besides the taxonomic significance, many species of the genus are also well-known for their medicinal properties which can be further investigated for proper utilization. I am sure, the data on morphology and ecology provided in the present work will be quite useful in constructing a broad phylogenetic classification of the species within the genus. I congratulate the author of the book Dr. L. B. Chaudhary for bringing a detailed taxonomic account on a very complex group of plants *Astragalus*.

Saroj K. Barik
Director
CSIR-National Botanical Research Institute
Lucknow

Phones: (Off.) (0522) 2205848, EPABX Phone: (0522) 2297800 to 2298999 Fax: (0522) 2205839,
Pox Box No. 436, Lucknow-226001
e-mail: director@nbri.res.in, sarojkbarik@gmail.com, Website: http://www.nbri.res.in

Contents

List of Photos

Introduction

Astragalus L. belongs to subfamily Papilionoideae of family Legumonosae under tribe *Galegeae* (Polhill, 1981; Lock and Schrire, 2005). It is the largest genus in angiosperm with about 2900 species, out of which 2400 species are distributed in old world and 500 species in new world (Podlech & Zarre, 2013; Zarre & Azani, 2013). It occurs throughout the world mostly in cold to warm arid and semiarid mountainous regions. The plants are chiefly perennial rhizomatous or hemicriptophytic herbs (2500 spp.) to cushion forming spiny shrubs (300 spp.) and to annual herbs (80 spp.) (Zarre & Azani, 2013). Podlech & Zarre (2013) have recently divided the entire old world species into 136 sections which have been gouped into three categories, 1. Annual species (21 sections), 2. Perennial species with simple basifixed hairs (75 sections) and 3. Perennial species with bifurcate hairs (40 sections). Since the genus has huge number of species and exhibits great morphological variability, the taxonomy of *Astragalus* is very challenging and large number of work has been carried out to solve the phylogeny at generic, subgeneric and sectional levels. The molecular data has shown that Astragalean clade (*Astragalus* L. and other closely related genera) are well supported monophyletic group and within Astragalean clade, except some outlier species (*i.e. A. camplanatus* R. Brown, *A. sinicus* L., *A. vogelii* (Webb) Bornm., *A. epiglottis* L., *A. cysticalyx* Ledeb. and *A. plecinus* (L.) Barneby) all *Astragalus* L. species are also monophyletic (Kazempour Osaloo *et al.*, 2003, 2005; Wijciechowski, 2005). However, the molecular data do not support subgeneric as well as sectional classification. Chaudhary *et al.* (2008) and Zarre & Azani (2013) have summarised the work so far carried out on the genus.

Since last more than 15 years I am working on the genus in India. Based on several field tours and examination of large number of herbarium specimens housed at BM (partially) BSD, BSHC, CAL, CDRI, DD, K (partially), KATH, LBG, LWG, P (partially) and TUH, 79 species have been recorded here from present political boundary of India in *Astragalus* L. Nine species have been excluded as they do not occur in India. A list of such species has been provided at the end. Among the studied species, 14 species are endemic to India. In India, the species are majorly distributed in the Himalaya at an altitudes ranging from (2000–) 3000–6000 m, except a few annual ones which are occasionally found in the plains of Haryana, Punjab and Rajasthan (Figure 1). The maximum concentration of the species lies in dry cold desert region of the Western Himalaya than more humid and moist Eastern

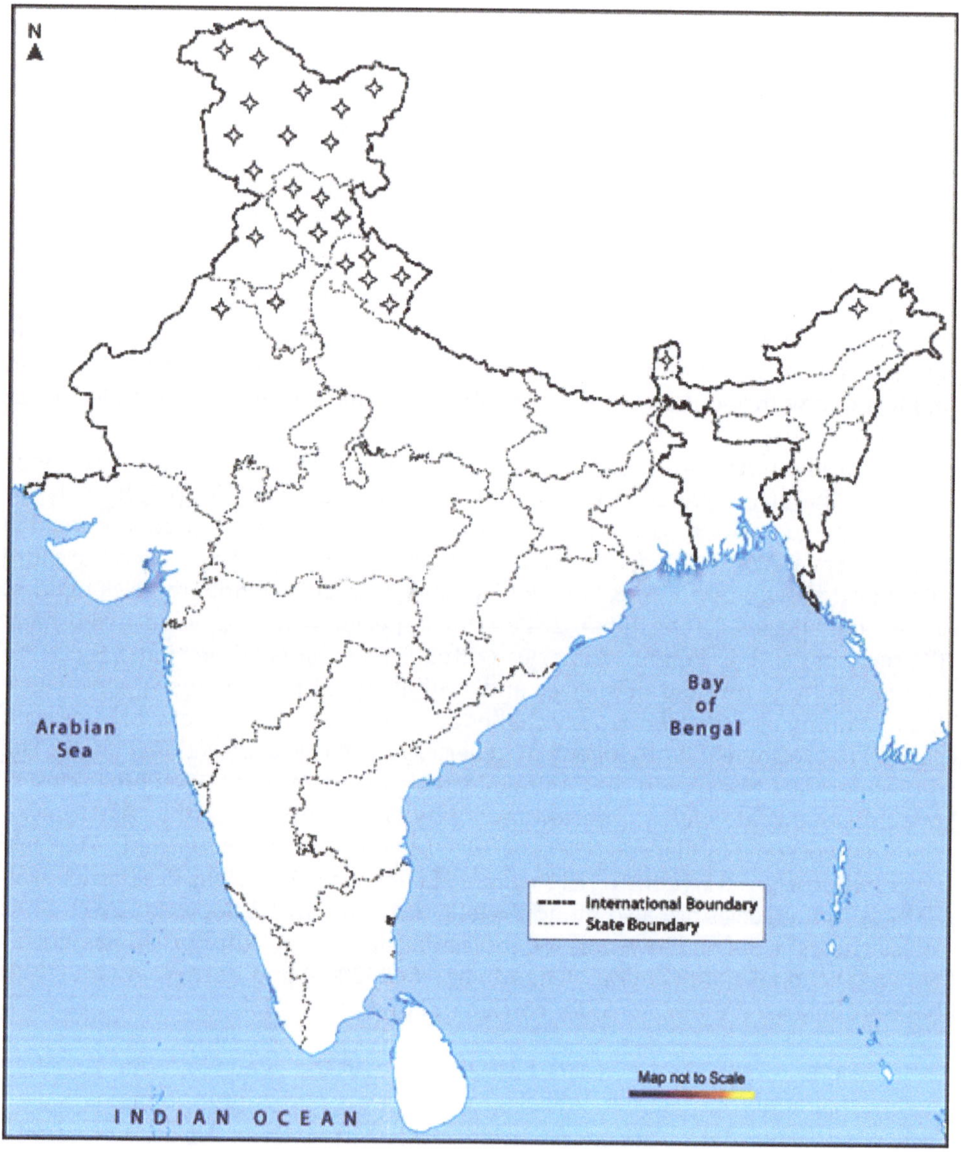

Figure 1. ✧ Distribution of *Astragalus* L. in India.

Himalaya. As per the recent classification, the Indian species belong to 28 sections. The distribution of all species in different sections has been shown in table 1.

Majority of the species of *Astragalus* L. grow in very hostile conditions of the cold deserts in the Himalaya. Generally they develop very thick and long rhizomatous rootstock, prostrate, caespitose and cushion forming habit, dense hairs, minute leaflets and rigid and spiny leaf rachis to counter the environmental stress. The species complete their life cycle (flowering and fruiting) within a short span of time usually during July to August (–September) because the habitats where they grow are covered with snow for most of the time. Although, due to variations in the habitats, the species exhibit tremendous variations in morphological characters, however, life span (annual or perennial), hairs (present or absent, basifixed or medifixed, appressed, ascending or spreading, white, black or brown, terete or ribbon like), habit (prostrate, erect or cushion forming), stems (present or absent, solid or hollow, branched or unbranched), stipules (adnate to the petiole, connate or completely free from petiole as well as from each other, connate in front of petiole or opposite the petiole, adnate or connate throughout or only in lower portion), leaves (imparipinnately or paripinnately compound), petiole and rachis (herbaceous or hard and spiny), leaflets (flat or folded, glabrous or hairy, covered with hairs on both sides or only on one side), inflorescence (peduncled or sessile, raceme, capitate-raceme or head, few or many-flowered), peduncle (shorter or longer, shorter, equal or longer than subtending leaf, elongating in fruits or remain unchanged in fruits), bracts (persistent or deciduous, longer or shorter than flower buds), bracteoles (present or absent, situated on pedicel or on calyx), flower colour (light yellow, yellow, purple, pink or white), pedicel (present or absent), calyx (tubular or campanulate, remain unchanged in fruits or become inflated in fruits, ruptured by fruits or remain unruptured, oblique or truncate at mouth , tube equal to or shorter than teeth, teeth equal or unequal, filiform or triangular, inside glabrous or hairy), petals (equal, subequal or unequal to each other), stamens (diadelphous or monadelphous, staminal tube obtuse, truncate or emarginate at mouth), ovary (stipitate or sessile, glabrous or hairy, one or many-ovuled), style (glabrous throughout or hairy below), stigma (glabrous or hairy), pods (linear, oblong, ovoid or ellipsoid, small or large, flat or turgid, grooved or obtuse dorsally, beaked at apex or without a beak, included into the calyx or excluded from the calyx, glabrous or hairy, unilocular, partially bilocualr or completely bilocularor) and seeds (rounded or kidney-shaped, turgid or flat) play important role in segregation of species/taxa. The discussion on these characters has been provided elaborately by Chaudhary *et al.* (2008) and Podlech & Zarre (2013).

Based on morphological (i. e. structure of pollen and wing and keel petals) and molecular features, Kang *et al.* (2003) and Kazempour Osaloo *et al.* (2005) have treated *Astragalus* sect. *Phyllolobium* (Fisch.) Bunge as a separate genus and therefore the species belonging to this section *A. donianus* DC., *A. heydei* Baker, *A. lachungensis* L. B. Chaudhary and *A. tribulifolius* Bunge have been treated under the genus *Phyllolobium* Fisch. following the recent treatments of Zhang & Podlech (2006) and Podlech (2008). Since people may search these species in *Astragalus* L. due to their long placement in the genus, the genus *Phyllolobium* Fisch. is also treated in the present work at the end. Similarly, *Astragalus aegacanthoides* Parker, *A. hosackioides* (Royle ex Benth.) Benth. and *A. zemuensis* W.W. Smith are also now treated under *Caragana aegacanthoides* (Parker) L.B.Chaudhary & S.K. Srivast., *Podolotus hosackioides* Royle ex Benth. and *Oxytropis zemuensis* (W.W. Smith) L.B.Chaudhary respectively.

In the present work the sections and species within the section have been arranged alphabetically. The keys to the sections and species have been provided for easy and quick identification. The treatment of sections is basically based on Podlech & Zarre (2013). The description of species has been exclusively prepared based on the Indian materials. Only relevant synonyms pertaining to the Indian flora have been provided. The measurement of pods excludes stipe and beak length and flowers exclude pedicel length.

The author is grateful to the Director, CSIR-National Botanical Research Institure, Lucknow for providing all necessary facities to accomplish this work. I am highly thankful to my teacher and mentor Dr. M. Sanjappa, Ex Director, Botanical Survey of India, Kolkata for his valuable suggestions on the subject. I express my sincere gratitude to the Indian Botanical Liaison Officers of Royal Botanic Gardens, Kew such as Drs. S. K. Srivastava, P. Lakshminarasimhan and V. Sampath Kumar for their support in sending valuable literature and information on Indian materials. My all colleagues and research scholars who helped me directly or indirectly in my research endeavour are duly acknowledged. Thanks are also due to the inchrge of various herbaria like BM, BSD, BSHC, CAL, CDRI, DD, K, KATH, LBG, LWG, P and TUH for granting permission to consult the herbarium and lending specimens for study. The Department of Science and technology (DST), Government of India, New Delhi and Department of Biotechnology, Government of India, New Delhi very generously funded the projects to carry out the research work on the genus. Last but not the least, my wife Mrs. Shweta Chaudhary and children Preksha and Onkar really deserve high appreciation for constant support and encouragement. Finally, the author sincerely thanks to Siya Publishing House, New Delhi for publishing this book.

Table 1. Sectional assignment of *Astragalus* species in India

Sl. No.	Section	Species
1	*Aegacantha*	*A. grahamianus, A. langtangensis, A. multiceps, A. oplites, A. polycanthus, A. psilocentros, A. webbianus, A. zanskarensis.*
2	*Ankylotus*	*A. commixtus, A. gracilipes.*
3	*Brachycarpus*	*A. heterodontus, A. karakorensis, A. melanostachys, A. sherriffii, A. thomsonii, A. tulinovii.*
4	*Caprini*	*A. candolleanus, A. drasianus, A. gilgitensis, A. malacophyllus, A. pindreensis, A. rhizanthus.*
5	*Cenantrum*	*A. floridulus, A. lessertiodes, A. pseudochlorostachys, A. tecti-mundi* subsp. *tecti-mundi, A. tecti-mundi* subsp. *orientalis.*
6	*Chaetodon*	*A. breviscapus.*
7	*Coluteocarpus*	*A. coluteocarpus* subsp. *colouteocarpus, A. coluteocarpus* subsp. *chitralensis.*
8	*Dissitiflori*	*A. imitensis, A. anfractuosus.*
9	*Galegiformes*	*A. chlorostachys, A. concretus, A. emodi, A. graveolens, A. isabellae, A. khasianus, A. maxwellii, A. nainitalensis, A. stewartii, A. stipulatus, A. trichocarpus, A. uttaranchalensis.*
10	*Helmia*	*A. arnoldii.*
11	*Herpocaulos*	*A. vogelii* subsp. *fatimensis.*
12	*Hispiduli*	*A. bakaliensis, A. scorpiurus.*
13	*Hookeriana*	*A. acaulis.*
14	*Hypoglottidei*	*A. gooraiensis, A. tibetanus.*
15	*Hypsophilus*	*A. nivalis.*
16	*Irinaea*	*A. amherstianus.*
17	*Komaroviella*	*A. alpinus, A. confertus, A. himachalensis, A. himalayanus, A. oxyodon.*
18	*Lithophilus*	*A. kashmirensis, A. munroi.*
19	Malacothrix	*A. pseudo-beckii.*
20	*Ophiocarpus*	*A. ophiocarpus.*
21	*Oroboidei*	*A. macropterus.*
22	*Poliothrix*	*A. cobresiiphilus, A. leucocephalus, A. minutidentatus, A. sanjappae, A. sikkimensis, A. strictus, A. tenuicaulis.*
23	*Pseudosesbanella*	*A. falconeri* var. *falconeri, A. falconeri* var. *pilosus, A. hoffmeisteri.*
24	*Pseudotapinodes*	*A. rigidulus.*
25	*Rhacophorus*	*A. verus.*
26	*Sesamei*	*A. tribuloides.*
27	*Skythropos*	*A. kongrensis, A. yunnanensis.*
28	*Uliginosi*	*A. peduncularis.*

Taxonomic Treatment

Astragalus

L., Sp. Pl. 755. 1753; Bunge, Mém. Acad. Imp. Sci. Saint Pétersbourg 11 (16): 1. 1868 & 15(1): 1. 1869; Baker in Hook. f., Fl. Brit. India 2: 118. 1876; Ali, Biologia 7: 7. 1961; Barneby, Mem. New York Bot. Gard. 13: 1. 1964; Ali in Nasir & Ali, Fl. West Pakistan 100: 121. 1977; Sanjappa, Legum. India 84. 1992; Kumar & Sane, Legum. South Asia: Checkl. 221. 2003; Lock & Schrire in Lewis *et al.*, Legum. World 481. 2005; L.B.Chaudhary & S.K. Srivast., Taiwania 52(1): 25. 2007; L.B.Chaudhary *et al.*, Taiwania 53(4): 338. 2008; Xu & Podlech in Wu *et al.*, Fl. China 10: 328. 2010; Podlech & Zarre, Tax. Rev. Gen. *Astragalus* (Legum.) Old World 1–3: 3. 2013. –Lectotype species: *Astragalus christianus* L.

Phaca L., Sp. Pl. 755. 1753. –Type species: *Phaca alpina* L.

Tragacantha Mill., Gard. Dict. Abr., ed. 4, vol. 3. 1754. – Lectotype species: *Tragacantha massiliensis* Mill.

Astragaloides Medik., Vorles. Churpfalz. Phys.-Ocon. Ges. 2: 373. 1787. –Lectotype species: *Phaca frigida* Jacq.

Astracantha Podlech, Mitt. Bot. Staatss. Munchen 19: 4. 1983. –Type species: *Astracantha cretica* (Lam.) Podlech.

Plants annual or perennial, herbs, undershrubs or shrubs, prostrate to erect, glabrous to hairy, with fibrous to thick, long, woody root-stock. Hairs basifixed or medifixed, appressed to spreading, white or black. Stipules present, adnate to the petiole or free, generally connate opposite the petiole (behind the stem). Leaves paripinnately or imparipinnately compound, rarely verticillate or unifoliolate (not in Indian species); leaflets entire, without stipels. Inflorescence axillary, sessile or pedunculate, racemose or spicate umbellate, loosely to densely few to many-flowerd, rarely one-flowered. Bracts present. Bracteoles present or absent. Flowers violet, purple, white, light yellow or yellow, often changing during anthesis, sessile to pedicellate, erect or nodded. Calyx quite variable, tubular or campanulate, occasionally split longitudinally in fruits, sometimes inflated in fruit, often oblique at mouth, teeth 5, equal to subequal. Petals glabrous or rarely hairy on back side, falling after anthesis or persistent; standard/vexillum ovate, obovate, oblong, panduriform or sometimes appendiculate, straight or backwardly curved in upper part, generally emarginated at apex, with or without distinct claw; wing petals equal to or shorter than standard, auriculate; keel petals generally shorter than or sometimes equal or longer than wings petals, joined to each other on ventral/lower margins; wing petals and keel petals interlocking with bosses/spurs. Stamens diadelphous ((9+1), occasionally closed monadelphous; staminal sheath truncate, obtuse or sometimes emarginate at apex; vexillary filament free or rarely attached with staminal sheath; anthers uniform. Ovary sessile to stipitate, glabrous to hairy; style straight to incurved in upper part, glabrous or sometimes hairy either in lower portion or near the tip; stigma mostly terminal, glabrous or occasionally penicillate. Pods variable, sessile or stipitate, 2-valved, compressed to turgid, mostly keeled on ventral side, grooved or flat to rounded on dorsal side, rarely rounded on both sides, glabrous to hairy, unilocular or partially to completely bilocular by a septum arising from dorsal side, dehiscing at ventral and dorsal side or only at dorsal side. Seeds usually oblong-reniform, compressed to turgid, smooth or with rough surface, glabrous.

Distribution: c. 2900 species in cold to warm arid and semiarid mountainous regions of the Northern hemisphere and South America; most diversified in South-Western Asia, South-Central Asia and Western-North Ameraca; center of origin in South-Western and South-Central Asia and the Himalaya; 79 species in India.

Key to the Sections

1a. Annual herbs 2

1b. Perennial herbs or shrubs 6

2a. Hairs medifixed; petals slightly longer than calyx ... **11. sect. *Herpocaulos***

2b. Hairs basifixed (to asymmetrically medifixed in sect. *Hispiduli* on pods); petals distinctly longer than calyx 3

3a. Pods narrowly linear, terete, flexuose-curved, slightly constricted between seeds, unilocular, valves hairy to glabrous, hairs of single type ....**20. sect. *Ophiocarpus***

3b. Pods linear or narrowly oblong to narrowly ovoid, grooved dorsally, slightly to distinctly curved, not constricted between seeds, partially to fully bilocular, valves hairy, hairs of two types-appressed short and spreading long 4

4a. Inflorescence dense head; pods narrowly oblong to narrowly ovoid, dilated at base .... **26. sect. *Sesamei***

4b. Inflorescence loose to dense raceme; pods linear, narrowed towards base 5

5a. Standard petal dilated in middle; style without a ring of minute hairs below stigma; hairs of pod not inserted at tubercules .... **2. sect. *Ankylotus***

5b. Standard petal not dilated in middle; style with a ring of minute hairs below stigma; hairs of pod mostly inserted at tubercules .... **12. sect. *Hispiduli***

6a. Hairs of plants medifixed or asymmetrical bifurcate 7

6b. Hairs of plants basifixed or plants partly or fully glabrous 11

7a. Plants dwarf with quite reduced stems ....**10. sect. *Helmia***

7b. Plants well developed with distinctly elongated stems 8

8a. Stipules free from each other 9

8b. Stipules connate opposite the petiole 10

9a. Calyx tubular and not inflated after anthesis; pods excluded from calyx .... **8. sect. *Dissitiflori***

9b. Calyx initially tubular or obliquely gibbous at base and narrowed towards mouth, after anthesis becoming ovoid-inflated; pods included into calyx .... **6. sect. *Chaetodon***

10a. Racemes globose, few-flowered; calyx tubular, after anthesis elongating and becoming inflated like bladder in fruiting; pods completely included into inflated calyx, small .... **15. sect. *Hypsophilus***

10b. Racemes loose and elongated, many-flowered; calyx remain tubular throughout; pods distinctly excluded from calyx, elongated .... **28. sect. *Uliginosi***

11a. Plants spiny; leaves paripinnately compound, occasionally with a minute rudimentary terminal leaflet when young, rachis indurating and persistent 12

11b. Plants not spiny; leaves imparipinnately compound, rachis usually not indurating and persistent 13

12a. Stems profusely and densely branched; leaf rachis rigid, straight, spreading, sharply spiny at tip; calyx turbinate, lobed up to the base, densely hairy; claws of wings and keels adnate to staminal sheath; pods unilocular .... **25. sect. *Rhacophorus***

12b. Stems normally branched; leaf rachis hard and spiny at tip; calyx tubular, pubescent; claws of wings and keels free from staminal sheath; pods partially to completely bilocular ....**1. sect. *Aegacantha***

13a. Plants stemless or sometimes shortly stemed 14

13b. Plants with well developed stems or sometimes almost stemless (as in sect. *Caprini* & *Brachycarpus*) 17

14a. Stipules connate opposite the petiole (behind the stem) 15

14b. Stipules connate in front of petiole 16

15a. Pods bilocular ....**19. sect. *Malacothrix***

15b. Pods unilocular .... **27. sect. *Skythropos***

16a. Petals yellow; pods almost bilocular .... **13. sect. *Hookeriana***

16b. Petals purple-red; pods incompletely bilocular .... **24. sect. *Pseudotapinodes***

17a. Plants erect shrubs or herbs 18

17b. Plants prostrate to ascending herbs or occasionally erect subshrub (sect. *Pseudosesbanella*) 21

18a. Stigma ciliate .... **7. sect. *Coluteocarpus***

18b. Stigma glabrous 19

19a. Pods unilocular .... **5. sect. *Cenantrum***

19b. Pods partially to completely bilocular (sometimes unilocular in sect. *Oroboidei*) 20

20 a. Stems hollow; pods sessile to subsessile .... **21. sect. *Oroboidei***

20b. Stems solid; pods distinctly stipitate .... **9. sect. *Galegiformes***

21a. Pods small, globose to ovoid, mostly sessile .... **3. sect. *Brachycarpus***

21b. Pods variable 22

22a. Calyx ruptured by pods .... **4. sect. *Caprini***

22b. Calyx remain unruptured 23

23a. Stipules shortly adnate to petiole or free 24

23b. Stipules free from petiole, connate opposite the petiole 26

24a. Stigma hairy .... **23 sect. *Pseudosesbanella***

24b. Stigma glabrous 25

25a. Stems covered with basifixed to bifurcate hairs; petals almost of the same length; pods generally upcurved .... **16. sect. *Irinaea***

25b. Stems covered with basifixed long spreading hairs; petals unequal in length; pods not straight .... **18. sect. *Lithophilus***

26a. Wing petals shorter than keel petals .... **17. sect. *Komaroviella***

26b. Wing petals longer than keel petals 27

27a. Pods erect, generally bilocular (unilocular in *A. gooraiensis*) .... **14. sect. *Hypoglottidei***

27b. Pods pendulous, unilocular or rarely incomplete bilocular .... **22. sect. *Poliothrix***

1. *Astragalus* sect. *Aegacantha*

Bunge, Mém. Acad. Imp. Sci. Saint Pétersbourg 11 (16): 42. 1868; Podlech & Zarre, Tax. Rev. gen. *Astragalus* (Legum.) Old World 1: 754. 2013. –Lectotype species: *Astragalus lasiosemius* Boiss.

Perennial spiny shrubs or shrublets, mostly cushion-forming. Hairs basifixed, white or black. Stipules membranous, adnate to petiole, sometimes connate opposite the petiole. Leaves paripinnately compound, in some species caducous terminal leaflet present only in young leaves; rachis hard and spiny. Raceme sessile or pedunculate, 1–few-flowerd. Bracts membranous, narrowly triangular to subulate, one-nerved. Bracteoles present or absent. Calyx tubular, ruptured by pod. Petals glabrous or hairy (not in the Indian materials), yellow, free from staminal sheath. Staminal sheath mostly oblique at apex. Pods thick, coriaceous, obliquely ellipsoid, with keeled and curved ventral side and more or less straight dorsal side, partially to completely bilocular.

Key to the Species

1a. Stems quite elongated with distinct internodes 2

1b. Stems absent, reduced or short with indistinct internodes 3

2a. Plants prostrate, not bushy; stipules adnate to petiole at base and also connate on opposide side of petiole; calyx oblique at mouth .... **6. *A. psilocentros***

2b. Plants procumbent to erect or bushy; stipules adnate to the petiole at base but not connate; calyx not oblique at mouth .... **3. *A. multiceps***

3a. Standard petal panduriform .... **1. *A. grahamianus***

3b. Standard petal not panduriform 4

4a. Stipules adnate to petiole and connate both; leaf rachis subspiny at tip .... **7. *A. webbianus***

4b. Stipules adnate to petiole only; leaf rachis spiny at tip 5

5a. Stems glabrous; leaves imparipinnately compound; leaf rachis erect and almost parallel to stem .... **2. *A. langtangensis***

5b. Stems sparsely hairy; leaves imparipinnately compound (very rarely terminal leaflet is seen at maturity); leaf rachis spreading or erecto-patent 6

6a. Stipules widely embracing stem, amplexicaul, hairy only along margins; wing petals equal to standard; pods sparsely hairy ...... **5. *A. polycanthus***

6b. Stipules obliquely inserted at stem, semiamplexicaul, hairy along margins as well as on outer surface; wing petals shorter than standard; pods normal hairy 7

7a. Leaves 2.5–6 cm long; leaflets hairy on both sides .... **8. *A. zanskarensis***

7b. Leaves 7–10 cm long; leaflets glabrous above, hairy below.....**4. *A. oplites***

1. *Astragalus grahamianus* Royle ex Benth. in Royle, Illustr. Bot. Himal. Mount. 199. 1835; Rechinger, Biol. Skr. 9 (3): 91, f. 101 & 102. 1958; Ali, Biologia 7: 63. 1961 & in Nasir & Ali, Fl. W. Pakistan 100: 190. 1977; Deml, Boissiera 21: 100. 1972; Sanjappa, Legum. India 88. 1992; Podlech in Fl. Iranica 175: t. 11. 2001; Kumar & Sane, Legum. South Asia: Checkl. 229. 2003; Podlech & Zarre, Tax. Rev. gen. Astragalus (Legum.) Old World 1: 771. 2013; L.B.Chaudhary & Siwakoti, Int. J. Curr. Res. Biosci. Plant Biol. 2 (4): 36, f. 1. 2015. –Type: India, Hills surrounding Cashmere, Royle s. n. (holo LIV n. v.; iso K!).

A. cicerifolius Royle ex Fisch., Bull. Soc. Imp. Naturalistes Moscou 26 (2): 404. 1853; Baker in Hook. f., Fl. Brit. India 2: 134. 1876, p. p.; Parker, For. Fl. Punjab 141. 1924. –Type: India, without locality, Jacquemont 1143 (holo P n. v.).

A. daltonianus Bunge Mém. Acad. Imp. Sci. Saint Pétersbourg 11 (16): 44. 1868. –Type: India, Himalaya bor.-occ., 10000–12000 ft., Thomson s. n. (holo P!; iso C n. v., GOET n. v., L n. v.).

A. polycanthus auct. non Royle ex Benth. 1835; Baker in Hook. f., Fl. Brit. India 2: 134. 1876, p. p.

Plate 32

Plants perennial shrubs, spiny, procumbent to suberect, stems woody, thick, profusely branched, completely covered with persistent leaf rachis, glabrous or thinly pubescent with half appressed, forwardly oriented, white hairs on main stems and sericeous on branches and young stems, the internodes comparatively shorter on main stems and longer on branches, not covered with stipules. Stipules 4–6 × c. 2 mm, adnate to the petiole at base, encircling the stem at base (amplexicaule), membranous, triangular-lanceolate with acuminate tip, hairy along margins, glabrous or thinly pilose with white, spreading hairs outside, faintly veined. Leaves 2.5–5.5 cm long, paripinnately compound; petiole 6–10 mm long, flattened at base; rachis 1.5–4.7 cm long, persistent, stiff, spiny at tip, grooved below or almost terete at maturity, pubescent to pilose with white, appressed or mixed appressed and spreading hairs; leaflets 4–7 pairs, 2–5 × 1–3.5 mm, subopposite to opposite, minutely petiolulate, deciduous, elliptic, oblong or obovate to narrowly obovate, cuneate at base, entire along margins, obtuse, acute, retuse or mucronate at apex, glabrous or sometimes glabrescent above, unevenly pubescent with appressed, white hairs below, hairs chiefly confined along margins and midrib region. Inflorescence 2–3-flowered raceme on very minute peduncle or solitary flower in the axil of leaf; rachis c. 6 mm long, pilose with mixed appressed and spreading white hairs. Bracts (1.5–) 2–4 mm long, persistent, linear, membranous, thinly hairy along margins. Bracteoles may or may not be present, 1, minute, situated at the base or just above the base of pedicel. Flowers yellow, 25–30 mm long; pedicel 2–4 (–6) mm long, pubescent with white appressed or half spreading hairs. Calyx 11–14 mm long, persistent, split when pods get matured, tubular, oblique at apex, tube 7–10 mm long, thinly pubescent with minute, appressed, white hairs outside, glabrous inside, lobes (2–) 2.5–4 mm long, triangular with acuminate tip, pubescent on both sides. Standard 25–30 × 10–11 mm, obovate with emarginate apex and narrow base, panduriform; wing petals 23–26 mm long, shorter than standard and longer than keel petals, lamina c. 9 × 4 mm, oblong, acute at apex, upper auricle c. 1 mm long, spurs present, claw 14–17 mm long; keel petals 21–22 mm long, shorter than wing petals, lamina c. 6 × 4.5 mm, incurved towards apex, upper auricle c. 1 mm long, spurs present, claw c. 15–16 mm long. Staminal sheath c. 21 mm long, free filaments 2–3 mm long; vexillary filament c. 22 mm long. Ovary c. 8–12 mm long, stipitate, pubescent, stipe c. 2 mm long; style 10–12 mm long, incurved in upper portion, pubescent in lower portion and glabrous in upper portion; stigma capitate, glabrous. Pods 11–14 × 4–6 mm, oblong, turgid, grooved dorsally, pointed at apex, stipitate (stipe c. 2 mm long), thinly pubescent with minute, appressed white hairs, faintly veined, partially to completely bilocular, c. 16-seeded.

Flowering & Fruiting: April–August.

Distribution: INDIA: Himachal Pradesh, Jammu & Kashmir; NEPAL; PAKISTAN.

Specimens examined: **Himachal Pradesh**: Chamba, Kilar, 2625 m, 18.7.1964, *Nair* 32559 (BSD). **Jammu & Kashmir**: Srinagar, 1219–1829 m, April–May 1936, *Timins* 45 (BM); Sind Valley, Kangan, 1676 m, 6.8.1940, *Ludlow & Sheriff* 7896 (BM); Kashmir, Hill sides, 2133 m, 1915, *Prescott-Decie s.n.* (BM); Kishtwar, 1524 m, 9.5.1976, *Stainton* 7543 (BM); Around Kishtwar, 25.5.1986, *Uniyal* 80263 (BSD); Hirpur, 1700 m, 9.6.1959, *Rao* 9367 (BSD).

Notes: The report of this species from Uttar Pradesh and Punjab is erroneous.

2. ***Astragalus langtangensis*** Podlech, Sendtnera 6: 136. 1999; Kumar & Sane, Legum. South Asia: Checkl. 233. 2003; L.B.Chaudhary & S.K.Srivast., Taiwania 52 (1): 28, f. 5. 2007; Podlech & Zarre, Tax. Rev. gen. Astragalus (Legum.) Old World 1: 776. 2013; L.B.Chaudhary & Siwakoti, Int. J. Curr. Res. Biosci. Plant Biol. 2 (4): 37, f. 1C. 2015. –Type: Nepal, Langtang valley, 13000 ft., 24.6.1965, Schilling, Sayers, Bista 380 (holo K!; photo LWG!; iso KATH!).

Plate 33

Plants perennial shrubs, spiny, up to 30 cm tall, stems densely covered with remnants of old stipules, glabrous, in upper part branched, branches erect, very densely leafy. Stipules 7–10 mm long, membranous, linear to narrowly triangular, adnate to the petiole in lower part, otherwise free, ciliate along margins. Leaves imparipinnately compound, 3.5–7 cm long; petiole 1–1.5 cm long; rachis erect, almost parallel to stem, slender but rigid, pungent, brown-blackish, glabrescent to glabrous, sometimes without leaflets; leaflets 9–11 pairs, 3–4 × 2–2.5 mm, obovate, truncate to emarginate, minutely mucronulate at apex, glabrous above, sparsely hairy below, terminal leaflet minute, caducous. Inflorescence 2-flowered raceme; peduncle c. 5 mm long. Bracts 4–5 mm long, membranous, linear-acute, white and black hairy. Flowers yellow; pedicel 3–4 mm long, erect, white and black hiary. Calyx 11–12 mm long, membranous, tubular, loosely to densely covered with spreading white and black hairs, teeth subulate, 3–4 mm long. Standard c. 22 × 7 mm, narrowly elliptic, emarginate at apex, gradually narrowed towards base into a long claw; wing petals c. 20 mm long, lamina c. 8 × 3 mm, narrowly gibbous-elliptic, obtuse at apex, auricle c. 1 mm long, claw 11–12 mm long; keel petals c. 17 mm long; lamina obliquely elliptic, curved in upper portion, subacute, auricle indistinct, claw c. 10 mm long. Ovary subsessile, white hairy; style hairy only in basal part. Pods 10–13 × 5–6 mm, subsessile, obliquely elliptic, flat to slightly and widely depressed dorsally, subabruptly narrowed into a 5–8 mm long slender beak at apex, bilocular, thinly transversely rugose, loosely covered with appressed to nearly spreading white hairs.

Flowering & Fruiting: October.

Distribution: INDIA: Uttarakhand-Kumaon (Rilkote- Upper Johar, Jola); NEPAL.

Specimens examined: Nil.

Notes: Specimens were not observed in the Indian herbaria. Description has been prepared after the study of type specimens and from Podlech & Zarre (2013).

3. ***Astragalus multiceps*** Wall. ex Benth. in Royle, Illustr. Bot. Himal. Mount. 199. 1835; Baker in Hook.f., Fl. Brit. India 2: 134. 1876; Deml, Biossiera 21: 169. 1972; Sanjappa, Legum. India 85.1992; Kumar & Sane, Legum. South Asia: Checkl. 234. 2003; Podlech & Zarre, Tax. Rev. gen. *Astragalus* (Legum.) Old World 1: 785. 2013; L.B.Chaudhary & Siwakoti, Int. J. Curr. Res. Biosci. Plant Biol. 2 (4): 38. 2015. –Type: India, Kunawur, *Royle s. n.* (lecto K!; iso CAL!, DD!, LE *n. v.*).

A. bicuspis Fisch., Symb. Astrag. Trag. 94. 1853; Baker in Hook. f., Fl. Brit. India 2: 135. 1876; Parker, For. Fl. Punjab 142. 1924; Ali, Biologia 7: 65. 1961 & in Nasir & Ali, Fl. W. Pakistan 100: 192. 1977; Deml, Biossiera 21: 82. 1972; Sanjappa, Legum. India 85.1992. –Type: India, between Lapau and Poya (upper Kunawur), *Jacquemont* 1695 (lecto LE *n. v.*; iso: K!, P!).

A. jacquemonti Bunge, Mém. Acad. Imp. Sci. Saint Pétersbourg 11 (16): 44. 1868. –Type: India, Himalaya bor.-occ., *Jacquemont* 1640 (lecto P *n. v.*; iso LE *n. v.*, P!).

?*A. leptocentrus* Bunge, Mém. Acad. Imp. Sci. Saint Pétersbourg 11 (16): 44. 1868; Baker in Hook. f., Fl. Brit. India 2: 135. 1876. –Type: in regno Cabulico, *Griffith s. n.* (Holo *n. v.*).

A. polycanthus auct. non Royle ex Benth. 1835; Baker in Hook. f., Fl. Brit. India 2: 134. 1876, p. p.

A. grahamianus auct. non Royle ex Benth. 1835; Fisch., Bull. Soc. Imp. Naturalisters Moscow 26 (2): 406. 1853.

Plates 1, 34

Plants perennial, spiny shrubs, procumbent to erect or bushy, profusely branched, stems quite elongated, reddish, almost terete, faintly striate, pilose, hairs white, thin, spreading. Stipules 4–6 × c. 1 mm, persistent, adnate to the petiole only at base, more or less membranous, narrowly lanceolate, more narrow and elongated at tip, sparsely ciliate with spreading hairs along margins, glabrous on both sides. Leaves 2.5–3.5 cm long, paripinnately compound; petiole 6–10 mm long; rachis 2–2.5 cm long, erecto-patent, almost terete, smooth, with 2– 5.5 mm long spiny tip, hairy as stem; leaflets 4–6 pairs, 4–12 × 1.5–5.5 mm, deciduous, minutely petiolulate, alternate, subopposite or opposite, oblong, oblong-elliptic, elliptic, oblong-ovate, obovate or narrowly obovate, cuneate at base, entire and sparsely ciliate with spreading hairs along margins, retuse, subtruncate, obtuse or acute with mucronate at apex, glabrous above, pilose below with thin, white, spreading hairs, sometimes hairs confined only on midrib on lower surface. Inflorescence axillary, (1–) 2–3-flowered on a very minute peduncle. Bracts 2–3.5 mm long, shorter than or equal to pedicel, membranous, persistent, linear-lanceolate, ciliate along margins with spreading hairs. Flowers 16–17 mm long, yellow; pedicel 2.5–3 mm long, pilose, hairs mixed black & white, spreading. Calyx 7.5–8.5 mm long, persistent, tubular, 5-lobed, sparsely pilose with spreading, mixed black & white hairs, tube c. 5 mm long, spliting on dorsal side in fruits, teeth 2–3.5 mm long, unequal, linear, shorter on dorsal side. Standard 16–17 × c. 8 mm, obovate/spatulate, emarginate at apex, claw not distinctly demarcated from lamina; wing petals 14–15 mm long, shorter than standard, lamina c. 7 × 2.5 mm oblong, obtuse at apex, upper auricles c. 1.2 mm long, claw c, 9 mm long; keel petals c. 13 mm long, shorter than wing petals, lamina c. 4.5 × 3 mm, oblong, curved above, obtuse at tip, upper auricle minute, spurs present, claw c. 9 mm long. Staminal sheath c. 9 mm long, obtuse at apex, free filaments unequal, 1.5–3 mm long, shorter and longer alternately arranged, thread like; vexillary filaments c. 12 mm long, free from staminal sheath. Ovary c. 6 × 1 mm, linear-elliptic, subsessile, densely sericeous; style c. 9 mm long, curved near tip, persistent, glabrous; stigma capitate, glabrous. Pods 10–13, subsessile, oblong, turgid, slightly grooved dorsally, pilose with thin, white, spreading hairs, prominently reticulately veined, c. 7-seeded, half bilocular. Seeds 2.5–3 × 2–2.5 mm, variable in shape, usually heart-shaped or reniform, glabrous, smooth, blackish-brown with scattered black spots.

Flowering & Fruiting: July–August (–September).

Habitat. Growing on dry mountain slopes in open or on flat lands near agriculture fields in hard or loose dry soils.

Distribution: INDIA: Himachal Pradesh, Jammu & Kashmir; NEPAL; PAKISTAN.

Specimens examined: **Himachal Pradesh**: Lahul Spiti: between Sissu & Gondla, 2900 m, 8.8.1970, *Bhattacharyya* 40696 (BSD); Sissu-Gondla, 3400 m, 16.7.1977, *Aswal* 6744 (CDRI); Gondla, 3200 m, 14.8.1982, *Aswal* 10927 (CDRI); Near Gondla, 3200 m, 3.8.2002, *Chaudhary & Khan* 225010 (LWG); Sissu-Gondla, 3400 m, 6.7.1973, *Bhattacharyya* 51882 (BSD); Between Sissu to Gondla, 18.7.2008, Chaudhary *et al.* 249608 (LWG); Thsirot, 6.8.1994, *Murti & Singh* 88129 (BSD); Tandi, 3100 m, 6.8.1971, *Bhattacharyya* 45138 (BSD); Tandi, 3100 m, 6.8.1971, *Bhattacharyya* 45141 (BSD); From Tandi to above, Mongla, 3500 m, 15.9.2001, *Chaudhary* 220671 (LWG); Gamur, 3300 m, 9.7.1973, *Bhattacharyya* 51918 (BSD); Between Keylong and Jispa , 3337 m, N 32° 35.374′ E 77° 07.640′, 29.7.2006, *Chaudhary et al.* 229427, 229428 (LWG); Lote, on way to Udaypur , 2931 m, N 32° 36.652′ E 77° 55.349′, 30.7.2006, *Chaudhary et al.* 229471(LWG); Near Jhalna, 2742 m, N 32° 40.688′ E 76° 42.331′, 30.7.2006, *Chaudhary et al.* 229482 (LWG); Lahul, collector? 260 (CAL). Without locality, 3048–3657 m, *Thomson s. n.* (CAL).

Notes: It is a very prominent and distinct species that occurs commonly in Lahul region (Lahul-Spiti) of Himachal Pradesh. It can easily be distinguished from other species by its bushy and spiny habit and elongated stems without crowded nodes. This species exhibits remarkable variations in the leaflets even on the same plant.

I could not observe any specimen from Jammu & Kashmir region. Podlech & Zarre (2013) have seen only one specimen from Rupshu at LE. The report from Uttarakhand is erroneous. Podlech & Zarre (2013) have not recorded this species from Nepal, while Chaudhary & Siwakoti (2015) have reported it from Nepal.

4. *Astragalus oplites* Benth. ex Parker, Kew Bull. 1921: 270. 1921 & For. Fl. Punjab 143. 1924; Ali, Biologia 7: 68. 1961 & in Nasir & Ali, Fl. W. Pakistan 100: 193. 1977; Sanjappa, Legum. India 92. 1992; Kumar & Sane, Legum. South Asia: Checkl. 235. 2003; Podlech & Zarre, Tax. Rev. gen. *Astragalus* (Legum.) Old World 1: 789. 2013; L.B.Chaudhary & Siwakoti, Int. J. Curr. Res. Biosci. Plant Biol. 2 (4): 38. 2015. –Type: in Tibeto-occ., 10000–14000 ft., *Thomson s. n.* (lecto LE *n. v.*; iso K!, P *n. v.*, STU *n. v.*).

A. zanskarensis Benth. ex Bunge subsp. *oplites* (Benth. ex Parker) Deml, Boissiera 21: 168. 1972.

A. cicerifolius Royle ex Bunge, Mém. Acad. Imp. Sci. Saint Pétersbourg 11 (16): 44. 1868, non Fisch. 1853.

Plates 2, 35

Plants perennial, spiny shrubs, erect, tufted, with short to well developed stems, internodes smaller than stipules, minutely pubescent with white hairs. Stipules 5–8 × 2–3 mm, persistent, adnate to the petiole at base, not encircling the stem, membranous, triangular-lanceolate, acuminate at apex, distinctly veined, minutely pubescent with white hairs. Leaves 7–10 cm long, paripinnately compound (due to early fall off apical leaflet all leaves appear paripinnate); rachis slender, stiff, persistent, erecto-patent, faintly grooved above, spiny at tip, pubescent with minute, white, appressed hairs; leaflets 8–14 apirs, 2–6 × 1.5–2.5 mm, deciduous, subsessile, obovate to narrowly obovate, emarginate, obtuse or subobtuse at apex, pilose with white, appressed hairs below, glabrous or unevenly glabrescent above. Inflorescence axillary, shortly peduncled raceme, 3–5-flowered; peduncle up to 15 mm long, sparsely pubescent with white, appressed, forwardly oriented hairs. Bracts 2–2.5 mm long, persistent, setaceous. Flowers c. 21 mm long; pedicel c. 3.5 mm long, pubescent with white and appressed hairs. Calyx c. 8 mm long, persistent, tubular, split in fruits, 5-lobed, tube c. 6 mm long, lobes unequal, upper ones c. 1 mm long, lower ones c. 2 mm long, pubescent outside, glabrous inside. Standard c. 21 × 4.5 mm, narrowly obovate with emarginate tip, narrow towards base, claw not very distinct; wing petals c. 19 mm long, shorter than standard and distinctly longer than keel petals, lamina c. 9 × 1–1.5 mm, oblong with subacute tip, upper auricle c. 1 mm long, spurs present, claw c. 11 mm long; keel petals c. 16 mm long, lamina c. 5 × 3.5 mm, upper auricle minute, spurs present, claw c. 11 mm long. Staminal sheath c. 14 mm long, obtuse at tip, free filaments unequal, 2–3 mm long, alternately longer and shorter; vexillary filament c. 16 mm long. Ovary c. 5 × 1.5 mm, hairy, stipitate (stipe c. 2 mm long); style 10 mm long, curved at tip, glabrous; stigma capitate, glabrous. Pods c. 10 × 4.5 mm, subsessile, oblong, turgid, slightly grooved dorsally, abruptly narrowed at apex, partially bilocualr, dehiscent, many-seeded, pilose with white hairs. Seeds c. 4 × 2.5 mm, turgid, oblong-reniform, brown, glabrous.

Flowering & Fruiting: July–August.

Habitat: Open dry sandy hill slopes, flat lands near agriculture fields, dry stony grounds and under *Salix* plantation. Altitudes 3200–5500 m.

Distribution: INDIA: Jammu & Kashmir, Himachal Pradesh, Uttarakhand; CHINA; NEPAL; PAKISTAN.

Specimens examined: **Himachal Pradesh**: Lahul-Spiti, Kibber, 4200 m, 18.9.2001, *Srivastava* 98886 (BSD); Kibber, 12.8.1994, *Murti & Singh* 88274 (BSD); Lossar, 4100 m, 26.7.1972, *Bhattacharyya* 48889 (BSD, CAL); 11 km before Kiato towards Lossar, 3900 m, 6.8.2002, *Chaudhary & Khan* 206782, 206783 (LWG); Tabo, 3300 m, 5.8.2002, *Chaudhary & Khan* 206780 (LWG); 1 km before Lossar, 4000 m, 4.8.2002, *Chaudhary & Khan* 206753, 206754 (LWG); Spiti vallley, before 2 km of Lossar from Kunjam pass, 4101 m, N 32° 26.858′E 76°43.455′, 1.8.2006, *Chaudhary et al.* 229511, 229512 (LWG); before 1 km of Lossar from Kunjam pass, 4097 m, N 32°26.164′E 77°44.456′, 2.8.2006, *Chaudhary et al.* 229515 (LWG); Lahul, Jispa, 3500 m, 15.8.1982, *Aswal* 10943 (CDRI); Spiti, Samdo-Kawrik, 4200 m, 21.8.1994, *Murti & Singh* 81594 (BSD); Spiti, Sogothang, 4100 m, 30.7.1972, *Bhattacharyya* 49147 (BSD); Spiti, Pin valley, 3810–3962 m, June year?, *Stolizcka s. n.* (CAL); Spiti, Thassigaong, 4572–4877 m, *Stolizcka s. n.* (CAL); Pangi, 3353–5029 m, 1879, *Heyde s. n.* (CAL); Lilang, around and above the village, 4572–4877 m, *Stolizcka s. n.* (CAL). **Jammu & Kashmir**: Ladakh, on way to Khardunga La, 3800 m, 9.8.1980, *Bhattacharyya* 71530 (BSD); Ladakh, Indus valley, 3400 m, 10.9.1970, *Bhattacharyya* 41110 (BSD); Upshi, 3500–3550 m, 6.8.1976, *Wadhwa* 59481 (BSD); Kurgiakh, 4300 m, 25.7.1973, *Bhattacharyya* 52239 (BSD, CAL); Ladakh, 20. 7. 1976, *Wadhwa* 58839 (BSD; Ladakh, 20.7.1976, *Wadhwa* 58839 (BSD); Ladakh, Leh, Sabu nullah, 3505 m, 1.8.1976, *Wadhwa*, 59355 (BSD); Leh, Phyang, 3800 m, 14.8.1975, *Vishwanathan* 54664 (BSD); Leh, 1–15. 07. 1856, *Schlagintweit s. n.* (BM); Leh, 3200 m, 23.8.1986, *Aswal & Goel* 16281 (CDRI); Karakoram, Sassir, 4328 m, July 1925, *Meinertzhagen s. n.* (BM); Baltistan, Haramosh, Barloogma nullah, 2743 m, 2.5.1928, *Ludlow* 251 (BM). **Uttarakhand**: Pithoragarh, on way to Garbyang, September 1986, *Malhotra & Balodi* 83384 (BSD).

Notes: Very close to *A. zanskarensis* Benth. ex Bunge and needs experimental study to establish correct status. The leaves generally appear paripinnate at maturity, however, imparipinnately compound condition has been observed in the present study specieally in young condition. The terminal leaflet is reduced and falls very early and therefore, leaves appear paripinnate. Podlech & Zarre (2013) have not recorded it from Nepal, however, Chaudhary & Siwakoti (2015) have seen it from Nepal.

5. *Astragalus polyacanthus* Royle ex Benth. in Royle, Illustr. Bot. Himal. Mount. 199. 1835; Baker in Hook.f., Fl. Brit. India 2:134. 1876 p. p.; Parker, For. Fl. Punjab 142. 1924; Ali, Biologia 7: 68. 1961 & in Nasir & Ali, Fl. W. Pakistan 100: 193. 1977; Deml, Boissiera 21: 142. 1972; Sanjappa, Legum. India 93. 1992; Kumar & Sane, Legum. South Asia: Checkl. 237. 2003; L.B.Chaudhary & S.K.Srivast.,Taiwania 52 (1): 29, f. 6. 2007; Podlech & Zarre, Tax. Rev. gen. *Astragalus* (Legum.) Old World 1: 793. 2013. –Type: India, Soognum in Kunawar, *Royle s. n.* (lecto LIV *n. v.*; iso K!).

A. tenuispinus Bunge, Mém. Acad. Imp. Sci. Saint Pétersbourg 11 (16): 44. 1868. –Type: India, Caschmiria *et* Kunaur, *Jacquemont* 1543 (lecto P *n. v.*; iso K!, LE *n. v.*, P *n. v.*).

Plants perennial shrubs, up to 15 cm long, with quite reduced internodes, glabrescent. Stipules c. 5–9 mm long, yellowish brown, membranous with several parallel nerves, triangular, adnate to petiole, widely clasping stem, amplexicaul, sparsely ciliate along margins, otherwise glabrous. Leaves paripinnately compound, 3.5–6.5 cm long, erect, crowded; petiole 1.5–2.5 cm long, hairy, glabrous with age; rachis spiny at tip; leaflets 10–12 pairs, 3.5–5 × c. 1.5 mm, always folded, elliptic or narrowly obovate, acuminate to truncate or shortly apiculate-mucronate, glabrous above, sparsely hairy at midrib and along margins below. Inflorescence 1–2-flowered raceme, axillary; peduncle 2.5–5 mm long. Bracts c. 3.5 mm long, narrowly triangular, ciliate along margins. Flowers yellow, c. 20 mm long; pedicel 2.5–4 mm long, sparsely hairy. Calyx 10–11 mm long, tubular, black and white hairy, lobes 2–3 mm long, subulate. Standard c. 20 × 8 mm, glabrous, without distinct claw; wing petals nearly as long as the standard, lamina c. 10 mm long, narrowly oblong, claw c. 10 mm long; keel petals c. 18 mm long. Staminal sheath oblique at mouth. Pods 15–17 × c. 4 mm, with a straight 3–5 mm long beak at tip, slightly grooved dorsally, sparsely hairy, c. 12-seeded.

Flowering & Fruiting: July

Distribution: INDIA: Himachal Pradesh, Jammu & Kashmir. Endemic.

Specimens examined: **Himachal Pradesh**: Kunawar, 1835, *Royle s. n.* (K). **Jammu & Kashmir**: Dras, 30.3.1922, *Stewart* 7402 (K).

Notes. Specimens of this species are not available in the Indian herbaria. Earlier it was reported from Nepal and Pakistan, but according to Podlech & Zarre (2013), it seems to be an endemic to India.

6. ***Astragalus psilocentros*** Fisch., Bull. Soc. Imp. Naturalistes Moscou 26 (2): 405. 1835; Parker, For. Fl. Punjab 140. 1924; Ali, Biologia 7: 62. 1961 & in Nasir & Ali, Fl. W. Pakistan 100: 188. 1977; Deml, Boissiera 21: 154. 1972; Sanjappa, Legum. India 93. 1992; Podlech in Fl. Iranica 175: t. 32. 2001; Kumar & Sane, Legum. South Asia: Checkl. 238. 2003; Podlech & Zarre, Tax. Rev. gen. *Astragalus* (Legum.) Old World 1: 796. 2013. –Type: Pakistan, Djillapour, Salt Range between Pind Dadan Khan and Jhelum, *Jacquemont* 109 (holo LE *n. v.*; iso K!, P!).

A. polemius Boiss., Fl. Or. 2: 306. 1872. –Type: Pakistan, Belutschia prope Zahri, *Stocks* 712 (holo G-BOIS *n. v.*; iso K *n. v*).

A. polyacanthus auct. non Royle ex Benth. 1835; Baker in Hook. f., Fl. Brit. India 2:134. 1876, p. p.

A. psilocentros var. *pilosus* Parker, For. Fl. Punjab 140. 1924. –Type: N.W. India, Normon on the Chenab, 1851, *Thomson s. n.* (lecto K!).

Plate 36

Plants prostrate, perennial spiny shrubs, c. 60 cm high, branched, branches terete, finely striate, glabrous below (older portion), thinly pilose with appressed, white hairs above (younger portion), internodes on main branches longer than stipules while on younger shoots quite reduced. Stipules c.5 × 1 mm, membranous, adnate to the petiole at extreme base, connate behind stem, linear-lanceolate, ciliate with white, long, hairs along margins and tip, faintly veined. Leaves 2–3.5 cm long, paripinnately compound (sometimes terminal leaflet present or it is very reduced and falls early), spreading or erecto-patent, very crowded on young branches; petiole 8–13 mm long, in hairs similar to rachis; rachis not much thick but stiff and spiny at tip, sparsely pilose with long, spreading, white hairs in young condition, glabrous at maturity; leaflets 3–4 pairs, 3–6 × 2–3 mm, opposite or subopposite, minutely stalked, obovate, elliptic or oblong-elliptic, acute, subobtuse or obtuse at tip, glabrous above, pilose with long, spreading, white hairs below. Bracts c. 2 mm long, linear, ciliate with white hairs. Flowers c. 25 mm long, axillary (on the young shoot), solitary or paired, yellow; pedicel c. 3 mm long, pilose with long spreading white hairs. Calyx 10–13 mm long, tubular, oblique at tip, pilose with white spreading hairs outside, glabrous inside, tube 8–10 mm long, lobes 2–3 mm long. Standard 23–25 mm long, obovate, emarginate at apex, claw not distinct; wing petals c. 23 mm long, slightly shorter than standard, lamina c. 6 × 3 mm, oblong-elliptic with obtuse tip, upper auricle c. 1 mm long, claw c. 17 mm long, spurs present; keel petals c. 22 mm long, slightly shorter than wing petals, lamina c. 6 × 4 mm, incurved towards apex, upper auricle minute, claw c. 16 mm long. Staminal sheath c. 20 mm long, obtuse at mouth, free filaments 2–3 mm long; vexillary filament 2–3 mm long. Ovary c. 6 mm long, subsessile, very densely sericeous; style c. 15 mm long, filiform, incurved in upper portions; stigma capitate, glabrous. Pods c. 12 × 5 mm, subsessile, turgid, oblong-elliptic, with c. 4 mm long beak, pilose with white spreading hairs, bilocular.

Flowering & Fruiting: March–May.

Distribution: INDIA: Jammu & Kashmir; AFGHANISTAN; PAKISTAN.

Specimens examined: **Jammu & Kashmir:** Udhampur, Jhajar Kotli, 2.4.1986, *Swami* 852 (BSD).

Notes: This species is very poorly represented in Indian herbaria. Only one specimen has been observed at BSD. Podlech & Zarre (2013) have not recorded it from India.

7. ***Astragalus webbianus*** Grah. (in Wall., Cat. 5936. 1831–32, *nom. nud*) ex Benth. in Royle, Illustr. Bot. Himal. Mount. 199. 1835; Baker in Hook. f., Fl. Brit. India

2: 132. 1876; Ali, Biologia 7: 55. 1961 & in Nasir & Ali, Fl. W. Pakistan 100: 156. 1977; Sanjappa, Legum. India 96. 1992; Kumar & Sane, Legum. South Asia: Checkl. 244. 2003; Podlech & Zarre, Tax. Rev. gen. *Astragalus* (Legum.) Old World 1: 651. 2013. –Type: India, Himachal Pradesh, Nako in Hungarung, *Royle s. n.* (lecto LIV *n. v*; iso K!, LE *n. v*).

A. minutifoliolatus Wendelbo, Nytt Mag. Bot. 1: 43. 1952. –Type: Pakistan, Chitral, Barum Gol glacier, 4500 m, 3.7.1950 *Wendelbo s. n.* (holo O *n. v*; iso BM *n. v*, K!, RAW *n. v*).

Plates 3, 37

Plants perennial subshrubs, subspiny, tufted, erect, caespitose, stemless or with short stems, with numerous branches arising from a thick woody rootstock, stems (if present) up to 12 cm long, woody, crowded with nodes, completely covered with stipules and leaf base. Stipules 10–18 × 3–4 mm, adnate to petiole at base, connate on back side of petiole, whitish-membranous, longitudinally multiveined, lanceolate, ciliate along margins, glabrous or sparsely pilose with white hairs outside, glabrous inside. Leaves 10–21 cm long, imparipinnately compound; petiole 3–9 cm long; rachis 5–18 cm long, erecto-patent, terete, indurated, stiff with subspiny tip, the petiole and rachis sparsely pilose with spreading, long, white hairs in young condition and glabrous at maturity; leaflets 15–27 pairs, alternate, subopposite or opposite, 4.5–10 × 3–5 mm, subsessile, deciduous, obovate, oblong, oblong-elliptic or narrowly ovate, cuneate at base, obtuse or retuse at apex, glabrous above, sparsely pilose with spreading, long, white hairs below in young condition and glabrous at maturity except midrib; petiolules less than 1 mm long, glabrous. Inflorescence axillary, short-peduncled; raceme (chiefly in the axils of upper leaves near tip of stem), 4–5-flowered; peduncle 5–10 mm long (in fruiting condition), glabrescent to sparsely pilose with spreading white hairs in lower portion and black hairs in upper portion. Bracts 8–10 mm long, longer than pedicel, persistent, whitish-membranous, linear, sparsely pilose with spreading, mixed black and white hairs or only ciliate with spreading white hairs. Flowers 20–25 mm long, yellow; pedicel c. 6 mm long, glabrous. Calyx 10–12 mm long, persistent, tubular, membranous, glabrous or with a few hairs on the margins of lobes, tube 7–9 mm long, teeth 3–4 mm long, linear. Standard 20–25 × c. 6 mm, obovate with retuse apex; wing petals 18–19 mm long, lamina 8–9 × c. 3 mm, oblong, obtuse at apex, upper auricle c. 1.5 × 1 mm, curved, claw 10–11 mm long; keel petals c. 15 mm long, shorter than wing petals, lamina c. 5 × 4 mm, claw c. 10 mm long. Staminal sheath c. 15 mm long, oblique at mouth, free filaments alternately shorter and longer; vexillary filament c. 16 mm long. Ovary c. 5 mm long, stipitate, c. 11-ovuled, glabrous or with few spreading, long, thin white hairs, stipe c. 2 mm long; style c. 10 mm long, incurved above, stigma capitate, glabrous. Pods 12–20 mm long, subsessile, oblong or oblong-elliptic, turgid, narrowed at both ends, with a 3–4 mm long straight beak, with membranous valves, not prominently veined, glabrous, bilocular, 11–20-seeded.

Flowering & Fruiting: July–August.

Habitat: Growing in rocky soil in open. Altitudes 3700–4877 m.

Distribution: INDIA: Himachal Pradesh; AFGHANISTAN; CHINA; PAKISTAN.

Specimens examined: **Himachal Pradesh:** Lahul-Spiti: Zingibar, 3800 m, 17.8.182, *Aswal* 10979 (CDRI); Zingibar, 3830 m, N 32° 47.262'E 77° 17.098', 29.7.2006, *Chaudhary et al.* 229444 (LWG); Zingibar, 4267 m, 15.7.1938, *Bor* 13985 (DD); Patscio, 3700 m, 24.8.1970, *Bhattacharyya* 40849 (BSD); Patscio, 4.8.2003, 3800 m, *Chaudhary & Khan* 225023 (LWG); Patscio, 3731 m, N 32° 40.315'E 77° 11.882', 29.7.2006, *Chaudhary et al.* 229440 (LWG); Near Patscio, 3785 m, N 32° 45.567' E 77° 15.602', 30.7.2011, *Chaudhary et al.* 263811 (LWG); Kenlung, 4572 m, 20.7.1941, *Bor* 16493 (DD); Yanansho, 4877 m, 2.7.1941, *Bor* 15142 (DD); Spiti, Pin Valley National Park, Larang La-Tariya, 4800 m, 6.8.2003, *Chandra Sekar* 103758 (BSD).

Notes: It is chiefly characterized by erecto-patent, terete, stiff with subspiny tip and glabrous leaf rachis, glabrous leaflets and glabrous and membranous pods. Podlech & Zarre (2013) have placed it in the sect. *Caprinii*, however, based on its stiff and subspiny nature of rachis, it has been assigned here to the sect. *Aegacantha*.

8. ***Astragalus zanskarensis*** Benth. ex Bunge, Mém. Acad. Imp. Sci. Saint Pétersbourg 11 (16): 43. 1868 & 15 (1): 67 1869; Baker in Hook. f., Fl. Brit. India 2: 134. 1876; Ali , Biologia 7: 58. 1961 & in Nasir & Ali, Fl. W. Pakistan 100: 171. 1977; Sanjappa, Legum. India 96. 1992; Kumar & Sane, Legum. South Asia: Checkl. 245. 2003; Podlech & Zarre, Tax. Rev. gen. *Astragalus* (Legum.) Old World 1: 803. 2013. –Type: India, Himalaya occidental-borealis alpinis, proper Zanskar, 10000–14000 ft., *Thomson s. n.* (lecto P *n. v.*; iso CAL!, CGE *n. v.*, GOET *n. v.*, K!, L *n. v.*, M *n. v.*, MSB *n. v.*, W *n. v.*).

Plates 4, 38

Plants perennial, spiny shrubs, tufted, suberect to erect, up to 5 cm long, deep rooted, densely branched, cushion forming, main stems and branchlets short, woody with densely crowded nodes (internodes very short), completely covered with stipules and persistent leaf base. Stipules 5–7 × 1–2 mm, adnate to petiole at base on back, membranous, lanceolate, pilose with white hairs along margins and on upper surface, glabrous on lower surface or sometimes hairs present only along margins, prominently multi-veined. Leaves 2.5–6 (–10) cm long, paripinnately compound (due to early fall of reduced terminal leaflet all leaves appear paripinnate), petiole 10–20 mm long, flattened at base; rachis greenish or reddish, persistent, stiff, ascending, spiny at tip after the fall of terminal leaflet, petioles and rachis moderately pilose with white, half appressed hairs in young condition, glabrescent to glabrous at maturity or densely hairy with spreading hairs; leaflets 9–14 pairs, c. 4 × 1.5 mm, subopposite to opposite, sometimes alternate in lower portion, subsessile, thick, deciduous, obovate, cuneate at base, acute, obtuse, retuse or emarginate at apex, moderately pilose with white, spreading hairs below and glabrous to glabrescent above or densely sericeous on both sides with appressed or spreading hairs. Inflorescence axillary, 2–4-flowered raceme; peduncle minute to 10 mm long, pilose with white, half appressed or spreading hairs. Bracts 3.5–6 mm long, shorter than or ± equal to pedicel, membranous, linear-lanceolate, hairy outside. Flowers 18–23 mm long, yellow; pedicel 4–6 mm long, hairy as peduncle or hairs mixed black and white. Calyx 7–12 mm long, tubular, sparsely pilose to glabrescent with appressed or spreading, white or mixed white and black hairs, dorsal lobes 1–2 mm long, ventral lobes 2–2.5 mm long, all linear. Standard 18–23 × c. 6 mm, elliptic-obovate with emarginate tip, claw not distinct; wing petals 16–20 mm long, shorter than standard and longer than keel petals, lamina 9–10 × c. 2.5 mm, oblong, subobtuse at apex, claw 9–10 mm long; keel petals 13–17 mm long, lamina 6–8 × c. 3.5 mm, claw c. 9 mm long. Staminal sheath c. 12 mm long, free filaments arising from obtuse tip of staminal sheath, unequal, 1.5–3 mm long. Ovary c. 8 × 1 mm, pilose with white hairs, shortly stipitate; style c. 7 mm long, incurved in upper portion; stigma capitate, glabrous. Pods c. 9 × 3 mm, oblong, turgid, slightly grooved dorsally, narrowed at tip into a 2.5–5 mm long straight beak, shortly stipitate (stipe c.1 mm long), pilose or densely sericeous with white, appressed hairs, partially bilocular.

Flowering & Fruiting: June–August.

Habitat: Common on very dry hard soil on hill slopes and gravelly ground. Altitudes 3300–5200 m.

Distribution: INDIA: Himachal Pradesh, Jammu & Kashmir, Uttarakhand; NEPAL; PAKISTAN.

Specimens examined: **Himachal Pradesh**: Spiti, 6 km before Kiato towards Lossar side, 3950 m, 6.8.2002, *Chaudhary* & *Khan* 206785 & 206786 (LWG); 1 km before Lossar, 4000 m, 4.8.2002, *Chaudhary* & *Khan* 206755, 206756 (LWG); Pagmo, 7 km before Kaza, 3700 m, 4. 8. 2002, *Chaudhary* & *Khan* 206765 & 206766 (LWG). **Jammu & Kashmir:** Leh-Ladakh, Fotula Pass, 3105 m, N 34° 17.548′ E 76° 41.016′, 2.8.2011, *Chaudhary et. al.* 263844 (LWG); Karakoram, Dabadas, 3200 m, 27.8.1939, *Russell* 1764 (BM); Karakoram, Nagar state, Minapin glacier, 3505 m, 26.8.1961, *Lloyd* & *Megaw* 116 (BM); Ladakh, Yeru (after Upshi), 3550 m, 6.8.2003, *Chaudhary* & *Khan* 225042 (LWG); Leh, Phey, 3400 m, 7.8.2003, *Chaudhary* & *Khan* 225046 (LWG); Leh, between Leh & Khardunga La, 4200 m, 8.8.2003, *Chaudhary* & *Khan* 225055 (LWG); Ladakh, Mathoo Phu, 4500–5200 m, 7.8.1976, *Wadhwa* 59645 (BSD); Ladakh, Kimi, Upper Nubra, 3355 m, 26.7.1947, *Schomburg* 18 (BM); Lama Puru Pass, 3657 m,, 34°13′ N, 76°35′ E, 18.6.1977, *Stainton* 7812 (BM); Zanskar, Phirtse La, 4724 m, 33°3′ N, 77.°5′ E, 21.7.1976, *Stainton* 7704 (BM); Zanskar, Padam, 3353 m, 11.7.1981, *Southampton University Botanical Expedition* 7 (BM); Dras, 3048 m, 34°28′ N, 75°57 E′, 7.6.1977, *Stainton* 7804 (BM); Ganglas, 7.8.1988, *Chowdhery* & *Uniyal* 86055 (BSD); Kashmir, Rama above Aster, 3353 m, 27.7.1946, *Stewart* 22927 (BM). **Uttarakhand**: Chamoli, Geoldong proper, 5000 m, 16.9.1975, *Nathani* 56285 (BSD); Chamoli, Niti proper, 3800 m, 8.9.1975, *Nathani* 56123 (BSD). N. W. Himalaya: Almi Pass, 3353 m, 31.5.1891, *Lace* 2097 (CAL).

Notes: Ali (1977) has correctly mentioned that the leaves in *A. zanskarensis* are imparipinnately compoundly compound. The terminal leaflet is reduced and caducous, therefore the leaves appear paripinnate at maturity. The terminal leaflet can be observed only in very young leaves. Sometimes, it becomes so reduced that it is represented only by a minute linear structure.

The species shows quite variations in density and types of hairs. In some plants, the hairs are dense and spreading, while in others they are thin and almost appressed. The colour of leaf rachis also varies from green to red. The length of flowers, bracts, pedicels and peduncles also shows considerable variations among different populations and even sometimes on the same plant.

2. *Astragalus* sect. *Ankylotus*

Bunge, Mém. Acad. Imp. Sci. Saint Pétersbourg 11 (16): 15. 1868; Podlech & Zarre, Tax. Rev. gen. *Astragalus* (Legum.) Old World 1: 128. 2013. –Type species: *Astragalus ankylotus* Fisch. & C.A.Mey.

Annual herbs. Hairs basifixed. Stipules almost free or shortly adnate to petiole. Racemes sessile or pedunculate. Calyx tubular. Petals unequal in length; standard dilated in the middle, abruptly narrowed towards tip into a tongue like appendix, claw present; wing petals incised or obtuse at tip. Pods sessile, linear, curved, deeply grooved dorsally, with a straight to curved to hooked beak, partially bilocular, valves covered with double indumentum of appressed short and spreading long hairs.

Key to the Species

1a. Stipules free; leaflets glabrous to unevenly pilose above, sparsely pilose below; pods with a hooked beak at apex, septum of pod extending more than half the length of pod .... **1. *A. commixtus***

1b. Stipules slightly adnate to petiole at base; leaflets densely pilose on both sides; pods with a straight beak at apex, septum of pod extending only up to half the length of pod ....**2. *A. gracilipes***

1. ***Astragalus commixtus*** Bunge, Arbeiten Naturf. Vereins Riga 1: 246. 1847; Podlech in Rech. f., Fl. Iranica 174: t. 20. 1999; Kumar & Sane, Legum. South Asia: Checkl. 225. 2003; Xu & Podlech in Wu *et al.*, Fl. China 10: 333, f. 355. 6–11. 2010; Podlech & Zarre, Tax. Rev. gen. *Astragalus* (Legum.) Old World 1: 129. 2013. –Type: Uzbekistan, zwischen Agatme and Karagata, 14.4.1842, *Lehmann s. n.* (lecto LE *n. v.*; iso G-BOIS *n. v.*).

A. affghanus Boiss., Fl. Or. 2: 1095. 1872, non Bunge 1868; Ali, Biologia 7: 30. 1961; Bot. Notis. 118: 91. 1965 & in Nasir & Ali, Fl. W. Pakistan 100: 166, f. 17 U–U1, 23 B. 1977.

A. brahuicus Boiss., Fl. Or. 2: 235. 1872, non Bunge 1868. –Type: Pakistan, in Belutschia, 1851, *Stocks* 716 (holo G-BOIS *n. v.*, iso K *n. v.*).

Plate 39

Plants annual herbs, procumbent, c. 16 cm long (including peduncle length), stemless or with short stems, stems if present 1.5–6 cm long, shorter stems densely pubescent with white, half appressed, forwardly oriented hairs and with internodes shorter than stipules, longer stems sparsely pubescent with white, appressed hairs and with internodes longer than stipules. Stipules c.1.5 mm long, free, linear-lanceolate, ciliate with white, long, spreading hairs. Leaves 4.5–7 cm long, imparipinnately compound; petiole 1.8–2.5 cm long; rachis 1.5–4.5 cm long, hairy as stem; leaflets 11–17, 5–9 × 2.5–3 mm, subsessile, opposite, thick, narrowly elliptic or narrowly obovate, retuse or truncate at apex, glabrous to unevenly thickly pilose above, sparsely pilose with white, long, spreading hairs below. Inflorescence axillary, peduncled capitate raceme, (1–) 2–4 (–5)-flowered; peduncle (in fruits) (2.5–) 5–11.5 cm long, glabrescent or thinly hairy with white, appressed hairs. Bracts minute, whitish membranous, white and ciliate. Flowers not seen. Pods 12–14 × 3–4 mm, sessile, cylindrical, deeply grooved dorsally, turgid, slightly to distinctly curved, abruptly curved at apex into a hooked beak, pubescent with white, mixed long and minute hairs, 10–12-seeded, half bilocular, septum extending more than half the length of pod, diminissing towards base and apex. Seeds c. 2.5 × 1.5 mm, reniform, turgid, brown, glabrous, with rugous surface.

Flowering & *Fruiting*: July–Aug.

Distribution: INDIA: Himachal Pradesh; CHINA; MIDDLE EAST; NORTHERN EURASIA; PAKISTAN; WEST ASIA.

Specimens examined: **Himachal Pradesh**: Spiti, Rangrik, 3800 m, 6.8.1972, *Bhattacharyya* 49355 (BSD).

Notes: New record to India. It is very close to *A. gracilipes* Benth. ex Bunge but differs in having pod septum extending more than half the length of the pod, while in *A. gracillipes* the septum is extending only up to the half the length of the pod.

2. *Astragalus gracilipes* Benth. ex Bunge, Mém. Acad. Imp. Sci. Saint Pétersbourg 11 (16): 15. 1868 & 15 (1): 17. 1869; Baker in Hook. f., Fl. Brit. India 2: 122. 1876: Ali, Biologia 7: 29, 1961 & in Nasir & Ali, Fl.W. Pakistan 100: 152, f. 17 T, T1, 23 I. 1977; Sanjappa, Legum. India 88.1992; Podlech in Rech. f., Fl. Iranica 174: t. 21. 1999; Kumar & Sane, Legum. South Asia: Checkl. 229. 2003; Podlech & Zarre, Tax. Rev. gen. *Astragalus* (Legum.) Old World 1: 131. 2013. –Type: Tibet Occid., 11000–14000 ft., *Thomson s. n.* (lecto P; iso BR *n. v.*, BREM *n. v.*, C *n. v.*, CGE *n. v.*, G *n. v.*, GOET *n. v.*, K!, L *n. v.*, LD *n. v.*, LE *n. v.*, MSB *n. v.*, P *n. v.*, STU *n. v.*, W *n. v.*, ZT *n. v.*, CAL!).

Plants annual suberect herbs, stemless or with very short stems, densely pilose with white, forwardly oriented, half spreading hairs, internodes shorter than stipules. Stipules minute, triangular, slightly adnate to petiole at base, pilose with white hairs. Leaves imparipinnately compound, 10–20 mm long; petiole and rachis densely pilose with white, forwardly oriented, half spreading hairs; leaflets 4–6 pairs, 2–4 × 0.5–1 mm, subsessile, oblong or narrowly obovate, obtuse at apex, densely pilose on both sides with white, spreading hairs. Inflorescence axillary, almost sessile to pedunculate capitate raceme (head), 1–3-flowered; peduncle 1–40 mm long, hairy as leaf rachis. Bracts minute, whitish-membranous, white ciliate. Flowers 5–7 mm long, subsessile, pale yellow. Calyx c. 4 mm long, tubular, persistent, densely pilose with white hairs outside, tube c. 3 mm long, lobes c. 1 mm long, setaceous. Standard c. 7 × 2 mm, blade in the middle dilated (appendiculate), narrowed towards base into a claw, abruptly narrowed into a tongue-like appendix in upper part with rounded to slightly retuse apex; wing petals c. 6 mm long, shorter than standard, lamina c. 3 mm long, narrowly oblong with obtuse tip, auricle minute, claw c. 3 mm long; keel petals c. 4 mm long, shorter than wing petals, lamina c. 1.5 × 1 mm, obovate-triangular, curved in upper part, obtuse at apex, auricle minute, claw c. 2.5 mm long. Staminal sheath c. 3 mm long, truncate at apex, free filaments c. 1.5 mm long; vexillary filament c. 3.5 mm long. Ovary c. 3 mm long, sessile, densely pilose with white hairs; style c. 2 mm long; stigma glabrous. Pods 7–15 × 3–4 mm, sessile, cylindrical, slightly curved, deeply grooved dorsally, abruptly curved at tip with a straight beak, densely pubescent with mixed short and long, white, forwardly oriented, half spreading hairs, 6–12-seeded, partially bilocular, septum exceeding only up to half the length of pod, less than 1 mm wide, diminissing towards base and apex. Seeds c. 2 × 1.5 mm, reniform, turgid, brown with black spots, glabrous, rugose.

Flowering & Fruiting: June–July (–August).

Distribution: INDIA: Jammu & Kashmir, Himachal Pradesh; AFGHANISTAN; CHINA, KAZAKHISTA; PAKISTAN.

Specimens examined: **Himachal Pradesh**: Lahul-Spiti, Spiti, Poh Gram, 3400 m, 5. 8. 2002, *Chaudhary & Khan* 206779 (LWG). **Jammu & Kashmir**: Ladakh, Fotula, 4267 m, July 1905, *Meebold* 1389 (CAL).

3. *Astragalus* sect. *Brachycarpus*

Boriss. in Komarov, Fl. U.R.S.S. 12: 877. 1946; Podlech & Zarre, Tax. Rev. gen. *Astragalus* (Legum.) Old World 2: 867. 2013. –Lectotype species: *Astragalus melanostachys Benth.*

Perennial herbs with mostly branched rootstock, prostrate to ascending, acaulescent to with well developed stems. Stipules shortly adnate to the petiole or free, distinctly connate opposite the petiole (behind stem), rarely very shortly connate or all free from each other in lower ones. Leaves imparipinnately compound. Racemes generally distinctly pedunculate, capitate, few to many-flowered. Flowers small, violet, purplish or yellow. Calyx campanulate to more rarely shortly tubular-campanulate. Petals glabrous, deciduous; standard mostly without distinct claw; wing petals shorter than standard, rounded or emarginated at apex; keel petals shorter than wing petals or all petals almost equal. Pods small, globose to ovoid, mostly sessile, bilocular or unilocular, 2–4-seeded.

Key to the Species

1a. All stipules free from petiole as well as from each other also; pods unilocular .... 3. *A. melanostachys*

1b. All stipules free from petiole but lower ones connate opposite the petiole and upper ones free from each other; pods partially or completely bilocular 2

2a. Wing petals emarginate at apex .... **2. *A. karakorensis***

2b. Wing petals obtuse at apex 3

3a. Stems glabrous or very sparsely hairy in upper part; leaflets glabrous on both sides or sparsely hairy on lower surface .... **6. *A. tulinovii***

3b. Stems hairy throughout; leaflets glabrous above, hairy below 4

4a. Standard widely elliptic to suborbicular, rounded to emarginate at apex; pods globose, rounded at apex without a beak, completely bilocular .... **1. *A. heterodontus***

4b. Standard elliptic to rhombic-elliptic, deeply and widely incised at apex; pods obtriangular, widely truncate at tip with persistent style, partially bilocular 5

5a. Leaflets linear; standard without a short claw; wing petals narrowly oblong with obtuse tip .... 4. ***A. sherriffii***

5b. Leaflets narrowly elliptic; standard abruptly contracted into a short claw; wing petals narrowly oblong and somewhat dilated towards truncate to rounded tip .... **5. *A. thomsonii***

1. *Astragalus heterodontus* Boriss., Trudy Tadzhikistansk. Bazy 2: 161. 1936; Gontsch., Flora Tadzhikistanica 5: 281, t. 27. 1937; Xu & Podlech in Wu *et al.*, Fl. China 10: 375. 2010; Podlech & Zarre, Tax. Rev. gen. *Astragalus* (Legum.) Old World 2: 876. 2013. –Type: Tadzhikistan, prope lacum Kara-kul, 2.7.1913, *Knorring s. n.* (lecto LE *n. v.*; iso LE *n. v.*).

Plants 10–25 cm tall, caespitose, stems slender, ascending, sparsely covered with very short appressed hairs, glabrescent with age. Stipules 2–4 mm long, ovate to triangular, ciliate along margins, lower ones shortly connate behind the stem, obtuse at apex, the upper ones nearly free from each other, acute at apex. Leaves 2–4 cm long; petiole 0.5–2 cm, like rachis loosely covered with appressed white and sometimes also black hairs; leaflets (4–)5–7(– 8) pairs, 5–12 × 3–4 mm, narrowly elliptic, obtuse at apex, sparsely appressed hairy below, glabrous above. Inflorescence 3.5–10 cm long, capitate to elongate raceme, densely many-flowered; peduncle 2–8 cm long, loosely covered with appressed black hairs in upper part. Bracts 1–2.5 mm long, membranous, linear, sparsely black hairy. Flowers violet (in dry condition). Calyx 2–3 (–4) mm long, campanulate, appressed black hairy, more rarely with some white hairs mixed in; teeth unequal, 1–1.5 mm long, 3 ventral subulate, close together, 2 dorsal triangular, remote. Standard 5–8 × 4–5 mm, widely elliptic to suborbicular, rounded to emarginated at apex, shortly cuneatly narrowed at base; wing petals 5–7 mm long, lamina 3.5–4 × 1.3–2 mm, narrowly oblong, obtuse at apex, auricle minute, claw c. 2 mm long; keel petals 4–5 mm long, lamina 2.5–3 mm long, obliquely elliptic curved, obtuse at apex, auricle minute, claw 1.2–2 mm long. Staminal tube truncate at apex. Pods 3–4 mm long, angled-globose, rounded at apex without beak, deeply grooved dorsally, bilocular; valves thin, cross-wrinkled, covered with spreading white and black hairs.

Flowering & Fruiting: July.

Distribution: INDIA: Uttarakhand-Kumaon; CHINA; TADZHIKISTAN.

Specimens examined: Nil.

Notes: Not examined in the present study. Recorded here after Podlech & Zarre (2013), who have reported it from India based on single specimen: Kumaon, Meerung, July 1844, *Munro s. n.* (K).

2. *Astragalus karakorensis* Pamp., Bull. Soc. Bot. Ital. 1915: 28. 1915; Podlech & Zarre, Tax. Rev. gen. *Astragalus* (Legum.) Old World 2: 879. 2013. –Type: India, Karakorum, passo di Karakorum, 5425 m, 1.8.1914, *Dainelli* & *Marinelli s. n.* (holo FI *n. v.*).

Plants 5–8 cm tall, stems 1–2 cm long, ascending, densely covered with subappressed to ascending white hairs. Stipules 4–5 mm long, free from petiole, ovate-triangular, acuminate at apex, ciliate along margins, lower ones obtuse at apex, often connate up to middle. Leaves 1.5–3 cm long; petiole 3–8 mm long, densely hairy like rachis; leaflets 6–8 pairs, 3–8 × 2–4 mm, ovate, obtuse at apex, glabrous on upper surface, densely covered with appressed to subappressed hairs on lower surface. Inflorescence raceme, densely many-flowered; peduncle 1.5–2.5 cm long, white hairy in lower portion, white and black hairy in upper portion. Bracts 3 mm long, membranous, narrowly triangular, white and black hairy. Flowers subsessile; pedicel minute, black hairy. Calyx c. 4 mm long, campanulate, oblique at mouth, densely covered with subappressed brown-blackis hairs, intermixed with some whitish hairs, teeth c. 1.5 mm long, unequal, the lower three subulate, the upper two narrowly triangular, all hairy on inner side. Standard c. 8 × 5–5.5 mm, strongly upcurved, suborbicular, deeply emarginated at apex, contrcted into a short claw at base; wing petals 6–6.5 mm long, lamina 4 × 2 mm, emarginated at apex, auricle minute, claw c. 2 mm long; keel petals c. 5 mm long, oblong, curved in upper part, obtuse at apex, auricle minute, claw c. 2 mm long. Staminal tube truncate at apex. Ovary subsessile, ellipsoid, white hairy; style short, glabrous.

Flowering & Fruiting: August.

Distribution: INDIA: Jammu & Kashmir-Karakorum. Endemic.

Specimens examined: Nil.

Notes: Not examined here. Recorded after Podlech & Zarre (2013) who have reported it from India based on collection 'Kashmir, Karakorum, passo di Karakorum, 5425 m, 1.8.1914, *Dainelli & Marinelli s. n.* (FI)'.

3. *Astragalus melanostachys* Benth. ex Bunge, Mém. Acad. Imp. Sci. Saint Pétersbourg 11 (16): 21. 1868; Baker in Hook.f., Fl. Brit. India 2: 125. 1876; Ali, Biologia 7: 33. 1961 & in Nasir & Ali, Fl. W. Pakistan 100: 159. 1977; Sanjappa, Legum. India 91.1992; Kumar & Sane, Legum. South Asia: Checkl. 234. 2003; Podlech in Fl. Iranica 178: t. 11. 2010; Podlech & Zarre, Tax. Rev. gen. *Astragalus* (Legum.) Old World 2: 883. 2013. –Type: tab. 5 in Klotzsch & Garcke, Bot. Erg. Waldem. Reise: 160 (neo).

A. bracteosus Klotzsch in Klotzsch & Garcke, Bot. Erg. Waldem. Reise: 160, tab. 5. 1862, non Boiss. & Noë 1856.

A. supraglaber Kitam., Acta Phytotax. Geobot. 16: 138. 1956. –Type: Afghanistan, Nuristan, Shabul Gul, 3900 m, 6.8.1955, *Kitamura s. n.* (holo KYO *n. v.*).

A. macrostegius Rech. f., Symb. Ahgh. 3: 37, t. 29. 30. 1957. –Type: Afghanistan, Nuristan, Minjan pass, 12000 ft., 26.7.1937, *Koelz* 12716 (lecto W *n. v.*; iso E *n. v.*, US *n. v.*).

Plates 5, 40

Plants perennial herbs, prostrate to suberect, 10–60 cm long, branched generally from the base, stems terete, striate, glabrescent below, sparsely pubescent above (younger portion), hairs minute, white throughout, blackish-brown at nodes, appressed. Stipules 7–10 × 5.5 –7 mm, free from petiole and from each other, foliaceous, ovate, acuminate at tip, sparsely white and black ciliate along margins, glabrescent outside, glabrous inside, faintly veined. Leaves 7–9.5 cm long, imparipinnately compound; petiole 1.5–2.5 cm long, with white or mixed black and white hairs, glabrescent with age, spreading; leaflets (8–) 10–11 pairs, 8–19 × 4–10 mm, opposite or subopposite, subsessile, ovate, narrowly ovate, obovate, narrowly obovate, oblong or elliptic, cuneate at base, sparsely ciliate with white hairs along margins, obtuse, truncate or retuse with a minute mucro at apex, glabrous both sides; petiolule minute, hairy. Inflorescence 9–16 cm long, axillary, peduncled capitate raceme; peduncle 6.5–14 cm long, terete, striate, sparsely strigose, hairs appressed, forwardly oriented, blackish-brown, dense towards tip; raceme 2.5–3 cm long, oblong, densely many-flowered. Bracts 8–9 × 2.5–3 mm, persistent, lanceolate, more acuminate at tip, sparsely to moderately hirsute outside, glabrous inside, hairs generally blackish-brown, spreading. Bracteole 1, c. 1 mm long, attached at the base of calyx, hirsute with blackish-brown and spreading hairs. Flowers c. 8 mm long, red or pale pink, subsessile. Calyx c. 5 mm long, persistent, campanulate, 5-lobed, tube c. 3.5 mm long, hirsute outside, glabrous inside, lobes subequal, 1.5–2 mm long, smaller on dorsal side, linear-triangular, hirsute both sides. Standard c. 8 mm long, lamina c. 4 × 3 mm, narrowly elliptic-obovate, retuse at apex, claw c. 4 mm long; wing petals c. 6.5 mm long, lamina c. 3 × 1 mm, obtuse and slightly recurved at tip, upper auricle c. 1 mm long, spurs present, claw c. 3.5 mm long; keel petals c. 6 mm long, lamina c. 2.5 × 2 mm, curved in upper portion, upper auricles and spurs present, claw c. 3.5 mm long. Staminal sheath c. 4.5 mm long, obtuse at apex, free filaments 1–1.5 mm long; vexillary filament c. 5 mm long; anthers minute, dorsifixed. Ovary c. 4 × 2 mm, elliptic-ovate, hirsute with black hairs; style c. 3.5 mm long, curved in upper part, glabrous; stigma capitate, glabrous. Pods c. 3.5 × 3 mm, included into calyx, sessile, ovoid, turgid, deeply grooved dorsally, distinctly beaked, unilocular, dehiscent, 2-seeded, strigose with black, appressed hairs (hairs ± spreading at tip portion). Seeds c. 2.7 × 1.5 mm, reniform, brown, glabrous, turgid, smooth.

Flowering & Fruiting: July–September.

Habitat: In sandy riverbeds, grassy moist hill slopes among rocks, and dry hard flat lands. Altitudes: 3350–5000 m.

Distribution: INDIA: Himachal Pradesh, Jammu & Kashmir, Uttarakhand; AFGHANISTAN; CHINA; NEPAL?; PAKISTAN; TADZHIKISTAN; UZBEKISTAN.

Specimens examined: **Himachal Pradesh**: Lahul-Spiti, Sissu, 3200 m, 25.9.1978, *Aswal* 10079 (CDRI); Khoksar, 3200 m, 23.8.1982, *Aswal* 11065 (CDRI); Koranchi, 12.7.1938, *Bor* 99 (DD); Pinkar, 3140 m, 19.7.1941, *Bor* 15492 (DD); Sissu, 2900 m, 7.8.1970, *Bhattacharyya* 40628 (BSD); Sissu, 3657 m, 6.7.1938, 12394 (DD); Lahul, Sissu, 2900 m, 7.8.1970, *Bhattacharyya* 40628 (BSD); Before 4 km of Chota Dara, 3391 m, N 32° 18.202′ E 77° 24.699′, 31.7.2006, *Chaudhary et al.* 229496 (LWG); Chhatru, 3400 m 13.9.1961, *Nair* 16943 (BSD); Shtitikar, 3950 m, 10.9.1961, *Nair* 16926 (BSD); Thango, 18.8.94, *Murti* & *Singh* 88473 (BSD); Chhatru, 3325 m, 13.9.1961, *Nair* 16926 (BSD); Chhatru, 3350 m, 13.9.1961, *Nair* 16937 (BSD); Chota Dara, 3475 m, N 32° 18.035′ E 77° 27.704′, 31.7.2006, *Chaudhary et al.* 229484 (LWG); Between Chota Dara and Batal, 3152 m, N 32° 17.654′ E 77° 31.297′, 1.8.2006, *Chaudhary et al.* 229504 (LWG); Chotadhara, 3850 m, 11. 9. 1961, *Nair* 16878 (BSD); 6 km before Chotadhara towards Khoksar, 4000 m, 6.8.2002, *Chaudhary* & *Khan* 206789 (LWG); Chotadhara, 3700 m, 3.8.2002, *Chaudhary* & *Khan* 206749 (LWG). **Jammu & Kashmir**: Ladakh, Sannodangsa, 3900–4300 m, 27.8.1976, *Wadhwa* 59899 (BSD); Bhotkal La, 3657 m, July 1905, *Meebold* 1391 (CAL). **Uttarakhand**: Tehri-Garhwal, Rhudughera, 3048–3353 m, 9.7.1883, *Duthie* 987 (DD); Tehri-Garhwal, water course near Gangla, 12. 07. 1883, *Duthie* 897 a (CAL); Uttarkashi, Bhojwasa to Gaumukh, 3700 m, 2.9.1983 *Bhattacharyya* 74769 (BSD).

Notes: Very distinct species by its foliaceous and free stipules, long peduncled capitate racemes, and sessile, 2-seeded and included pods into the calyx. Baker (1876) has observed oblong and half-biolocular pods, however, in the present study ovoid and unilocular pods have been observed like Ali (1977). The stipules are free from each other, however Ali (1977) and Podlech & Zarre (2013) have noticed connate stipules in this species. The species has been reported from Nepal by Chater (1979), Press *et al.* (2000) and Chaudhary & Siwakoti (2015), however, Podlech & Zarre (2013) have not recoreded it from Nepal.

4. ***Astragalus sherriffii*** Podlech, Sendtnera 7: 190. 2001; Kumar & Sane, Legum South Asia: Checkl. 240. 2003; L.B.Chaudhary & S.K.Srivast., Taiwania 52 (1): 29. 2007; L.B.Chaudhary *et al.*, Taiwania 52 (3): 222, f. 8 & 11E. 2007; Podlech & Zarre, Tax. Rev. gen. *Astragalus* (Legum.) Old World 2: 855. 2013. –Type: India, Kashmir, Ladakh, Gya, 13500 ft. 14.7.1941, *Ludlow* & *Sherriff* 8491 (holo E *n. v.*; iso BM!).

Plants perennial herbs, ascending to erect, c. 30 cm tall, stems slender, branched, several arising from base, sparsely pubescent with minute, appressed, white hairs. Stipules 3–3.5 × 1–1.5 mm, sparsely pilose with black or mixed black and white hairs outside, lower one ovate, free from petiole, connate on opposite side of petiole, upper one narrowly triangular, free from each other as well as from petiole. Leaves 3–6 cm long, imparipinnately compound; petiole 7–20 mm long, pubescent like stem; leaflets 3–5 pairs, 8–13 × 1.5–2 mm, subsessile, thick, linear-oblong or linear-lanceolate, subobtuse to obtuse at apex, entire along margins, glabrous above, pubescent with white, appressed hairs below; petiolule c. 1 mm long, pilose with half appressed, white hairs. Inflorescence 5–11 cm long, axillary, peduncled raceme; peduncle 4.5–8 cm long, pubescent with appressed, white and mixed black and white hairs especially towards tip; raceme 1–3 cm long, many-flowered, dense. Bracts c. 2 mm long, persistent, narrowly triangular, pilose with mixed black and white hairs. Flowers 7–8 mm long, suffused pink; pedicel minute, mixed black and white hairy. Calyx campanulae, c. 3 mm long, oblique at mouth, pilose with mixed black and white hairs; lobes unequal, upper two triangular, 0.5–1.0 mm long, lower three narrowly triangular, 1–1.5 mm long, hairy on inner side. Standard 7–8 × 4 mm, elliptic, retuse at apex, narrowed at base without a distinct claw; wing petals c. 6 mm long, lamina c. 4 × 1.5 mm long, narrowly obovate, obtuse at apex, claw c. 2 mm long; keel c. 4 mm long, lamina c. 2 × 2 mm long, obovate with incurved lower margin and straight to slightly concave upper margin, subobtuse at apex, claw c. 2 mm long. Staminal tube c. 3 mm long, truncate at the mouth. Ovary c. 2 mm long, sessile, glabrous; style c. 2.5 mm long, incurved towards tip; stigma capitate, glabrous . Pods not seen.

Flowering & Fruiting: July.

Distribution: INDIA: Jammu & Kashmir. Endemic.

Specimens examined: **Jammu & Kashmir:** Ladakh, Gya, 4115 m, 14.7.1941, *Ludlow & Sherriff* 8491 (BM).

Notes: Podlech & Zarre (2013) have observed wing petals with obtuse or retuse to emarginate tip in this species, however, in the present study only obtuse tip has been noticed. It is very close to *Astragalus thomsonii* Podlech, while Podlech & Zarre (2013) have placed it in a different section (i. e. *Oroboidei*).

5. ***Astragalus thomsonii*** Podlech, Sendtnera 6: 140. 1999; Kumar & Sane, Legum. South Asia: Checkl. 228. 2003; L.B.Chaudhary & S.K.Srivast., Taiwania 52(1): 30, f. 9. 2007; Podlech & Zarre, Tax. Rev. gen. *Astragalus* (Legum.) Old World 2: 894. 2013. –Type: Tibet occ., reg. alpina, 12000–15000 ft., *Thomson s. n.* (sub *A. densiflorus*) (holo MSB *n. v.*; iso K!, M *n. v.*, W *n. v.*).

Astragalus densiflorus auct. non Kar. & Kir. 1842; Baker in Hook. f., Fl. Brit. India 2: 125. 1876; Ali, Biologia 7: 32. 1961 & in Nasir & Ali, Fl. W. Pakistan 100: 161, f. 17 I. 1977; Sanjappa, Legum. India 94. 1992.

Plate 41

Plants perennial herbs, procumbent to erect, up to 30 cm long, with several stems from base, stems slender, finely striate, branched, glabrate to sparsely pubescent with minute appressed white hairs. Stipules 2–5 × 1–1.5 mm, free from petiole, connate in the lower portion of stem opposite petiole, almost to completely free in upper portion of stem, triangular with acute to acuminate tip and entire margins, pilose outside particularly along margins and tip with mixed black and white hairs, glabrous inside. Leaves 4–6 cm long, imparipinnately compound; petiole 2–20 mm long, pubescent like stem; rachis 20– 25 mm long, channeled above, pubescent like stem; leaflets 4–6 pairs, 8–16 × 1.5–3 mm, thick, shortly petiolulate, linear-oblong or linear-lanceolate, subobtuse to obtuse at apex, entire along margins, glabrous above, pubescent with appressed white hairs below; petiolules c. 1 mm long, pilose with half appressed white hairs. Inflorescene 5–11 cm long, longer than subtending leaf, axillary, peduncled raceme (appear like spike due to subsessile flowers); raceme many-flowered, dense; peduncle 4.5–8 cm long, pubescent with appressed white hairs, with mixed black and white hairs in upper portion; rachis pubescent with mixed white and black hairs. Bracts c. 2 mm long, longer than pedicel, persistent, linear-lanceolate, pilose with black or mixed white and black hairs. Flowers 7–8 mm long, sessile, purple or suffused pink, nodded; pedicel less than 1 mm long, pilose with black or mixed white and black hairs. Calyx c. 3 mm long, persistent, campanulate, oblique at mouth, pilose with black or mixed black and white hairs outside, glabrous inside except lobes, tube c. 2 mm long, lobes c. 1 mm long. Standard 7–8 × c. 4 mm, broadly elliptic-obovate with retuse apex, claw not very distinct; wing petals c. 6 mm long, shorter than standard, lamina c. 4 × 1.5 mm, oblong with obtuse tip, upper auricle and spurs present, claw c. 2 mm long; keel petals 4–4.5 mm long, lamina c. 2 × 2 mm, incurved in upper portion, upper auricles and spurs present, claw c. 2 mm long. Staminal sheath c. 3 mm long, almost with truncated apex, free filaments 0.5–1 mm long. Ovary c. 2 mm long, sessile, ellipsoid, turgid, glabrous; style incurved above, c. 2 mm long, glabrous; stigma capitate, glabrous. Pods 2–3 × 2–3 mm, sessile, obtriangular, turgid, cuneate at base, widely truncated at tip with persistent style, grooved dorsally, with rugose and faintly veined surface, pilose with white or mixed white and black hairs, partially septate by very narrow membrane (less than 1 mm wide) into two equal or unequal valves, 1–2-seeded. Seeds c. 2.5 × 2 mm, reniform, smooth, brown, glabrous.

Flowering & Fruiting: July–September.

Habitat: Altitudes 4000–5200 m.

Distribution: INDIA: Himachal Pradesh, Jammu & Kashmir; CHINA.

Specimens examined: **Jammu & Kashmir:** Ladakh, Tanglang La, 5200 m, 6.8.2003, *Chaudhary & Khan* 225037 (LWG); Ladakh, Ramshey, 3500 m, 21.8.1980, *Bhattacharyya* 71761 (BSD); Ladakh, Gomburang Zang, 4500 m, 27.7.1973, *Bhattacharyya* 52199 (BSD), Ladakh, Bangdum Gompa, 3962 m, July 1905, *Meebold* 1380 (CAL); Ladakh, Hemis National Park, way to Rumbak, 4000 m, 17.7.2002, *Srivastava & Chauhan* 100914 (BSD).

Notes: The true *Astragalus densiflorus* Kar. & Kir. in which stipules are adnate to petiole at base, ovary is hairy and pods are globose with rounded tip without beak does not occur in India. I could not examine the specimens from Himachal Pradesh. Podlech & Zarre (2013) have observed one specimen from Himachal Pradesh at K (*i. e.* Lahul, *Jaeschke* no. 39 of second collection).

6. ***Astragalus tulinovii*** O. Fedtsch., Fl. Pamir 27. 1903; Podlech & Zarre, Tax. Rev. gen. *Astragalus* (Legum.) Old World 2: 894. 2013. –Type: between Chodzha-Nasar and Pereval Koj-Tezek, 13000–14000 ft., 20.7. (flowering), *Fedtschenko s. n.* (lecto LE *n. v.*).

Plants 8–20 cm long, stems glabrous or in upper parts sometimes sparsely to loosely covered with ± appressed white and black hairs. Stipules 3–6 mm long, glabrous or with sparse black and white hairs at margins and apex, lower ones highly connate. Leaves 2–4 cm long; petiole 5–12 mm long, like rachis sparsely subappressed white and black hairy; leaflets 3–5 pairs, 5–8 × 1–3 mm, narrowly elliptic, obtuse to subacute at apex, glabrous or sparsely hairy on lower surface. Inflorescence 6.5–10.5 cm long, racemes, densely many-flowered; peduncle 5–8 cm long, loosely white hairy, towards tip increasingly black hairy. Bracts 4–5 mm long, sparsely black hairy. Flowers whitish. Calyx c. 5 mm long, densely black villous, teeth c. 2.5 mm long. Standard 7–8 × 4–5 mm, widely obovate to suborbicular, retuse at apex; wing petals 6–7 mm long; keel petals 4.5–5 mm long. Pods 3–4 × c. 3 mm, ovoid-globose, rounded at apex without beak but sometimes with persistent style, bilocular, 2-seeded, with smooth valves, covered with spreading white and black hairs.

Flowering & Fruiting : July–August.

Distribution: INDIA: Jammu & Kashmir-Ladakh; CHINA; PAKISTAN; TADZHIKISTAN.

Specimens examined: Nil.

Notes: Not examined here. Recorded after Podlech & Zarre (2013) who have reported it from India based on single collection 'Kashmir, Ladakh, NE slope of Tso Moriri, 5460 m, *Klimes* 99/11/6 (MBS)'.

4. *Astragalus* sect. *Caprini*

DC., Prodr. 2: 301. 1825; Podlech & Zarre, Tax. Rev. gen. *Astragalus* (Legum.) Old World 1: 463. 2013. –Type species: *Astragalus caprinus* L.

A. sect. *Myobroma* (Steven) Bunge, Mém. Acad. Imp. Sci. Saint Pétersbourg 11 (16): 33. 1868.

A. subgen. *Myobroma* (Steven) Baker in Hook. f., Fl. Brit. India 2: 118. 1876.

Myobroma Steven, Bull. Soc. Imp. Naturalistes Moscou 29: 150. 1856. –Type species: *Astragalus utriger* Pall.

Perennial herbs or subshrubs, acaulescent or with distinct stems. Hairs smooth, basifixed, white, sometimes also with black hairs chiefly in Inflorescence. Stipules adnate to the petiole, sometimes connate either opposite the petiole or in front of petiole, lower ones shorter, ovate with obtuse tip, glabrous, upper ones longer, narrowly triangular with acute tip, hairy. Leaves imparipinnately compound; rachis soft, rarely persistent. Racemes sessile or long pedunculate, 1–many-flowered. Bracts membranous, narrowly triangular to linear. Bracteloes rarely present. Flowers white or yellow. Calyx tubular, ruptured by pod. Petals glabrous; wing petals shorter than or equal to standard; keel petals shorter than wing petals. Ovary sessile or stipitate, glabrous or hairy; style glabrous or sometimes hairy in lower portion; stigma capitate, glabrous. Pods variable, sessile or stalked, with thin to coriaceous valves, unilocular or partially to completely bilocular.

Key to the Species

1a. Plants usually stemless or with short stems; stems without distinct internodes; flowers aggregate at the tip of stem 2

1b. Plants usually with well developed stems; stems with distinct internodes; flowers on the peduncle in the axil of leaves 5

2a. Ovary and pods glabrous or nearly so .... **2. *A. drasianus***

2b. Ovary and pods hairy 3

3a. Calyx teeth more or less equal to or longer than tube .... **3. *A. gilgitensis***

3b. Calyx teeth shorter than tube 4

4a. Leaf rachis neither persistent nor indurated; terminal and lateral leaflets not deciduous ....**6. *A. rhizanthus***

4b. Leaf rachis persistent, not indurated; terminal and lateral leaflets deciduous ....**5. *A. pindreensis***

5b. Plants densely hairy; leaflets hairy on both sides, terminal leaflet falls before lateral leaflets .... **4. *A. malacophyllus***

5a. Plants almost glabrous to moderately hairy; leaflets almost glabrous on upper surface and moderately hairy on lower surface, terminal leaflet does not fall before lateral leaflets .... **1. *A. candolleanus***

1. ***Astragalus candolleanus*** Royle ex Benth. in Royle, Illustr. Bot. Himal. Mount. 199. 1835; Baker in Hook.f., Fl. Brit. India 2: 132. 1876; Parker, For. Fl. Punjab 136. 1956; Ali, Biologia 7: 51. 1961 & in Nasir & Ali, Fl. W. Pakistan 100: 184. 1977; Sanjappa, Legum, India 85. 1992; Anand *et al.*, Taiwania 55 (3): 198, f. 1B. 2010. –Type: India, Kunawur, *Royle s. n.* (lecto K!).

A. rhizanthus subsp. *candolleanus* (Royle ex Benth.) Podlech, Mitt. Bot. Staatss. München 25: 184. 1988; Kumar & Sane, Legum. Soth Asia: Checkl. 239. 2003; Podlech & Zarre, Tax. Rev. gen. *Astragalus* (Legum.) Old World 1: 622. 2013.

Plates 6, 42

Plants perennial herbs, prostrate or semierect, with well developed stems, tufted, stems 10–25 cm long, branched, usually with condensed internodes covered with stipules and leaf rachis, thinly hairy. Stipules 15–30 × 7 mm, membranous, persistant, lanceolate or ovate with acute to quite elongated tip, prominently multi-veined, pilose outside or only ciliate along margins with white or mixed black and white hairs. Leaves 8–23 cm long, overtopping the stem, imparipinnately compound; petiole 4–11 cm long; rachis usually persistent, stiff, indurated, erecto patent, subspiny at tip after the fall of terminal leaflet, almost terete, thinly pilose with long spreading hairs; leaflets 10–13 pairs, 4–13 × 2–7 mm, gradually reducing in size towards apex and becomimg needle like terminally, deciduous, subsessile, opposite to subopposite, broadly elliptic, oblong or obovate to narrowly obovate, obtuse or retuse with a minute mucro or acute at apex, glabrous or sometimes glabrescent above, unevenly or throughout spreading pilose below. Inflorescence axillary, sometimes aggregate towards the tip of stem, subsessille to peduncled 3–8-flowered raceme or capitate raceme; peduncle 1.5–8.0 cm long, pilose. Bracts 8–18 × 1–3 mm, linear or lanceolate, pilose, ciliate along margins. Flowers 25–28 mm long, yellow or white; pedicel 3–4.5 mm long, hairy. Calyx 11–16 mm long, tubular, longitudinally splitting after the development of pod, thinly pilose with mixed black and white hairs or only with white or black hairs, tube 6–10 mm long, lobes (2.5–3.5–) 5–7 mm long, linear. Standard c. 25 mm long, emarginate at apex; wing petals c. 21 mm long; keel petals c. 16 mm long. Ovary c. 5 mm long, linear, densely hairy, stipe c. 1.5 mm long; style c. 10 mm long, abruptly incurved in upper portion, glabrous or sometimes hairy below; stigma capitate, glabrous. Pods 10–25 × c. 5 mm, sessile, oblong, narrowed towards base and apex into a long beak, turgid, grooved dorsally, thinly pilose with long spreading grey hairs, bilocular, dehiscent along dorsal suture.

Flowering & Fruiting: (May–June–) July–September.

Distrib: INDIA: Himachal Pradesh, Jammu & Kashmir, Uttarakhand; NEPAL; PAKISTAN.

Specimens examined: **Himachal Pradesh**: Kulu dist., Dibibokri Nal, 3900 m, 18. 6. 1952, *Schelpe* 3340 (BM); Spiti, Lossar, 4100 m, 25.7.1972, *Bhattacharya* 48865 (CAL); Lahul, Gondla, 2905 m, N 32° 36.286′ E 76°55.996′, 30.7.2006, *Chaudhary et al.* 229465 (LWG); Gondla, 3000 m, *Chaudhary* 220670 (LWG); Spiti, Lossar, 4100 m, 25 July 1972, *Bhattacharyya* 48865 (BSD); 1 km before Lossar, 4000 m, *Chaudhary* & *Khan* 206757 (LWG); Lossar, 4000 m, *Chaudhary* & *Khan* 206759 (LWG); Spiti vallley, before 1 km of Losar from Kunjam pass, 4106 m, N 32° 26.590′ E 76° 44.573′, 1.8. 2006, *Chaudhary et al.* 229506 (LWG); Before 1 km of Losar from Kunjam pass, 4106 m, N 32° 26.590′ E 76° 44.573′ 1.8. 2006, *Chaudhary et al.* 229508 (LWG); Losar , 4079 m, N 32° 25.848′ E 77° 55.0460′ 2.8. 2006, *Chaudhary et al.* 229517 (LWG); 1 km Before of Losar from Kunjam pass, 4088 m, N 32° 26.571′ E 76° 44.571′ 2.8. 2006, *Chaudhary et al.* 229521, 229522, 229523, 229524 (LWG); Baspa valley, Chitkul, hill slope on right bank of the river, above P.W.D. rest house, 24.9.1971, *Janardhanana*, 46194 (BSD); Baspa valley, Rakcham, 3000 m, 27.9.1964, *Nair*, 34341 (BSD); Rakcham, hill slope on left bank of river, c. 3300 m, 28.9.1971, *Janardhanan* 46335 (BSD); Chamba, 3048–3353 m, 23.8.1896, *Lace* 1464 (CAL); Branjen Wardwan, 2438 m, June 1905, *Meebold* 1370 p.p. (rightside specimen) (CAL). **Jammu & Kashmir**: Zanskar, Padam, 3353 m, 11.7.1981, *Southampton University Botanical expedition* 4 (BM); Pir Panjal, 3353 m, 6.7.1876, *Clarke* 28760 (BM); Kashmir, 2438 m, 1915, *Perscott-Decie s. n.* (BM); Kashmir, 1885, *Ledgnrik s.n.* (CAL); Sekiwas, 3353 m, 29.8.1956, *Polunin* 56/588 (BM); Baltal-Amarnath track, 3657 m, 15.7.77, 34°20′N, 75°28′E, *Stainton* 7950 (BM); Gadsar, 3962 m, 34°30′ N 75° 40′ E., 11.7.1977, *Stainton* 7913 (BM); Near Sonamarg, 3200 m, 16.8.1913, *Koebel* 138 (CAL); Sonamarg, 2743 m, 17.7.1928, *Stewart* 9719 (BM); Ladakh, Leh, Sabu Nullah, 3000 m, 1.8.1976, *Wadhwa* 59355 (BSD); Ladakh, collector ? 209b (CAL); Gilgit road, Deosai plains, Baltistan and Ladakh, Zojipass, Ladakh Road, 3353 m, 30.8.1940,

Stewart 21270 (BM); Leh-Ladakh, 15 km after Sonmarg towards Dras, 3420, N 34° 16.153'E 75°26.946', 6.8.2011, *Chaudhary et al.* 263893 (LWG); Leh-Ladakh, Zozila pass, 10 km after Sonmarg towards Dras, 3208 m, N 34°15.742'E 75°25.547', 6.8.2011, *Chaudhary et al.* 263888 (LWG); Ladakh, Dras, 27.7.1988, *Chowdhery & Uniyal* 85791 (BSD); Kashmir, Seshnag, 3600 m, 25.7.1966, *Nair*, 36944 (BSD); Masjid Gali near Tilel, 3353–3657 m, August 1939, *R. R. & I. D. Stewart s. n.* (BM); Synthen top, 4500 m, August 1986, *Vohra & Naithani*, 83144 (BSD); Sangla, 2591 m, 25.9.1964, *Brandis* 3799 p.p. (right side specimen) (CAL); without locality, 2743–3657 m, *Stoliczka* (CAL). Himalaya: Bor. Occ., date?, collector ?, 2438–3048 m, *Thomson s.n.* (CAL). **Uttarakhand**: Chamoli, opposite Gothing, 4200 m, 12. 9. 1975, *Naithani* 56200 (BSD); Soonderdunga valley, 4115 m, 30.5.1854, collector ? (CAL); Pithoragarh, on way to Relkot, June 1985, *Balodi* 77581 (BSD); Kumaon, Bogdwar/Martoli, 3000–4500 m, 14.6.1958, *Rao* 6789 (BSD); Kumaon, way to Ralam glacier, 4000–5000 m, 15.9.1969, *Pant & Naithani* 39699 (BSD); Kumaon, 35000–5000 m, 23.10.957, *Rao* 4436 (BSD).

Notes: Podlech & Zarre (2013) have treated this as a subspecies of *A. rhizanthus* Royle ex Benth. Although, it is very allied to *A. rhizanthus*, but can be differentiated in having less dense hair, elongated stems and pods with narrow both ends. Anand *et al.* (2010) have shown that it also differes from *A. rhizanthus* at molecular level.

2. *Astragalus drasianus* H.J.Chowdhery *et al.*, Bull. Bot. Surv. India 34: 209, f. 1. 1992 (publ. 1997); Kumar & Sane, Legum. South Asia: Checkl. 227. 2003; L.B.Chaudhary & Z.H.Khan, Phytotaxo. 4: 58, f. 2. 2004; L.B.Chaudhary *et al.*, Taiwania 52 (3): 217. 2007; Podlech & Zarre, Tax. Rev. gen. *Astragalus* (Legum.) Old World 1: 539. 2013. –Type: India, Jammu & Kashmir, Ladakh, Dras, *Chowdhery & Uniyal* 85804A (holo CAL!; iso 85804B BSD).

Plate 43

Plants perennial herbs, almost stemless. Stipules c. 15 mm long, membranous, narrowly lanceolate, adnate to the petiole at very base, glabrous or with some hairs along margins at tip, faintly veined. Leaves 19–25 cm long, imparipinnately compound; petiole 5–8 cm long, rachis neither terete nor stiff, sparsely pilose with spreading, long, white hairs; leaflets 22–27 pairs, 9–10 × 3–4 mm, alternate to opposite, oblong, oblong-elliptic, narrowly ovate or narrowly obovate, obtuse at tip, glabrous, long hairs found only on midrib. Inflorescence axillary; raceme c. 5-flowered; peduncle 5–10 mm long, glabrous. Bracts c. 8 mm long, longer than pedicel, persistent, linear, membranous, thinly ciliate along margins. Flowers c. 20 mm long; pedicel c. 6 mm long, glabrous. Calyx 10 12 mm long, tubular, glabrous or very thinly ciliate along the margins of lobes with long white hairs, tube c. 7 mm long, lobe linear, lower ones c. 4 mm long, upper one c. 3 mm long. Standard c. 20 × 6 mm, obovate, obtuse at apex, narrow towards base, claw not distinct; wing petals c. 20 mm long, shorter or equal to standard, lamina c. 9 × 3 mm, oblong, obtuse at tip, upper auricle c. 1.5 × 1 mm, claw c. 11 mm long; keel petals c. 15 mm long, shorter than standard and wing, lamina c. 5 × 4 mm, claw c. 10 mm long. Staminal sheath c.15 mm long; vexillary filament c. 16 mm long. Ovary c. 5 mm long, glabrous, c.11-ovuled, stipitate, stipe c. 2 mm long; style c. 10 mm long, incurved towards apex; stigma capitate, glabrous. Pods 17–23 × c. 10 mm, subsessile, with c. 5 mm long beak at tip, turgid, oblong or oblong-elliptic, prominently vein, glabrous or sometimes with a few thin, long, white, spreading hairs, bilocular.

Flowering & Fruiting: June–August.

Distribution: INDIA: Himachal Pradesh, Jammu & Kashmir. Endemic.

Specimens examined: **Himachal Pradesh**: Panji village, Bashahr, 7.6.1962, *Nair* 22537 (BSD). **Jammu & Kashmir**: Ladakh, Dras, 3200 m, 31. 5. 1941, *Ludlow & Sheriff* 8334 (BM); Near Dras town, 3091 m, N 34° 25.843′ E76° 44.693′, 5.8.2011, *Chaudhary et al.* 263882 (LWG).

Notes: It resembles with *A. rhizanthus* Royle ex Benth. but differs in having glabrous or nearly glabrous stipules, leaflets, pedicel, calyx, ovary and pods and wing petals equal to standard.

3. *Astragalus gilgitensis* Ali, Phyton 11: 135, f. 1 & 6. 1966 & in Nasir & Ali, Fl. W. Pakistan. 100: 183. 1977; Podlech, Mitt. Bot. Staatss. Munchen 25: 190, f. 22. 1988; Sanjappa, Legum. India 88. 1992; Kumar & Sane, Legum. South Asia: Checkl. 229. 2003; L.B.Chaudhary & S.K.Srivast., Taiwania 52 (1): 28, f. 4. 2007; Alam & Ali, Phyton 48 (2): 211.2009; Podlech & Zarre, Tax. Rev. gen. *Astragalus* (Legum.) Old World 1: 553. 2013. –Type: India, Gilgit, Nomal to Naltar, 17.7.1957, *Stewart* 26492 (holo RAW *n. v.*; iso K!).

Plants 15–20 cm longl, acaulescent, densely hairy. Stipules 10–14 mm long, membranous, narrowly triangular with acuminate tip, adnate to petiole, distinctly longitudinally nerved, densely covered with long, spreading hairs. Leaves imparipinnately compound, 27–31 cm long; petiole 1–5 cm long, densely covered with spreading hairs; leaflets 22–29 pairs, 6–10 × 3–6 mm, elliptic to ovate, subacuminate at apex, densely spreading hairy on both sides. Inflorescence axillary, sessile, raceme. Bracts 12–15 mm long, white-membranous, linear, acuminate at apex, hairy. Flowers yellow; pedicel 2–4 mm long, hairy. Calyx 18–19 mm long, tubular, densely villous, tube c. 9 mm long, teeth 9–10 mm long, filiform. Standard 18–21 mm long; lamian 8–10 mm long, obovate, emarginate at apex, gradually narrowed towards base into 10–11 mm long claw; wing petals 16–18 mm long, lamina 7–9 × 2.5–3 mm, narrowly oblong, rounded at apex, auricle c. 1 mm long, claw 9–10 mm long; keel petals c. 15 mm long, lamina c. 6 × 3–4 mm, obliquely elliptic-triangular, subacute, auricle indistinct, claw c. 9 mm long. Ovary 5–6 mm long, hairy, with c. 2 mm long stipe; style hairy up to the middle. Pods subsessile, c. 10 mm long, slightly grooved on dorsal side, with c. 3 mm long beak, partially bilocular, densely villous.

Flowering & Fruiting: July.

Distribution INDIA: Jammu & Kshmir-Gilgit; E. AFGHANISTAN; N. PAKISTAN.

Specimens examined: **Jammu & Kashmir**: Gilgit, Kattar to Nomal, 2438 m, 24.7.1954, *Stewart* 26535 (K); Gilgit, 3048 m, 18.7.1954, *Stewart* 26502 (K).

Notes. It is known only from Gilgit of Jammu & Kashmir (Ali, 1966; Podlech & Zarre 2013). Recently, Alam & Ali (2009) have treated this species as critically endangered. The specimens of the species are not available in the Indian herbaria. The species is very close to *A. rhizanthus* Royle ex Benth.

4. *Astragalus malacophyllus* Benth. ex Bunge in Mém. Acad. Imp. Sci. Saint Pétersbourg 11 (16): 36. 1868 & 12: 61. 1869; Baker in Hook. f., Fl. Brit. India 2: 133. 1876; Ali in Biologia 7: 54.1961 & in Nasir & Ali, Fl. W. Pakistan 100: 168, f. 17 N. 1977. Sangappa, Legum. India 91. 1992; Anand *et al.* in Taiwania 55 (3): 198, f. 1C. 2010 & Proc. Indian natn. Sci. Acad. 78 (2): 179, f. 3 & 4. 2012. –Type: India, in Himalayo bor.-occ., 8000–10000 ft., *Thomson* (lecto P!; iso BM *n. v.*, C *n. v*, CGE *n. v.*, E *n. v*, FI *n. v.*, G *n. v*, GOET *n. v.*, K!, L *n. v.*, M *n. v.*, STU *n. v.*, W *n. v.*).

Plates 7, 44

Plants perennial herbs with well developed stems and busy habit, caespitose, prostrate, stems up to 90 cm long with condensed to distinct internodes, branched, white tomentose. Stipules 6–18 x 5–3 mm, persistent, shorter to longer than internodes, adnate to petiole at base, lanceolate to narrowly lanceolate with acute to very long narrow tip, prominently longitudinally multi-veined, white pilose with spreading hairs outside, glabrous inside. Leaves 5.5–19 cm long, imparipinnately compound; petiole 2–3.5 cm long; rachis 8–15 cm long, persistent, subindurated, subspiny at tip after the fall of terminal leaflet, white pilose with spreading hairs; leaflets 18–21 pairs, close, 3–11 x 2–5 mm, subsessile, obovate, oblong, oblong-elliptic or almost orbicular, cuneate at the base, entire along margins, acute, subobtuse, obtuse or subtruncate at apex, tomentose on both surfaces, terminal leaflet smaller in size than lateral ones and sometimes reduced into needle like structure. Inflorescence axillary, shortly peduncled 3–5-flowered raceme; peduncle 2–2.5 cm long, mixed white and black (predominantly white) spreading hairs. Bracts 6–10 mm long, linear, longer than pedicel, sparsely pilose outside, glabrous inside, white ciliate along margins. Flowers 17–23 mm long, yellow; pedicel 4–5 mm long, mixed black & white tomentose. Calyx 12–14 mm long, persistent, tubular, mixed brown and white tomentose outside, glabrous inside, tube 7–10 mm long, longer than lobes, lobes 3–5 mm long, subequal, shorter on dorsal side, linear. Standard 17–22 mm long, deflexed, lamina c. 14 x 8 mm, obovate, emarginate at apex, claw c. 8 mm long; wing petals c. 21 mm long, more or less equal to standard, lamina c. 9 x 2.5 mm, oblong, obtuse at apex, upper auricle c. 1 mm long, spurs present, claw c. 12 mm long; keel petals c. 16 mm long, shorter than wing petals, lamina c. 5 x 3.5 mm, almost oblong, slightly incurved towards apex on lower margins, subacute at apex, upper auricle minute, spurs present, claw c. 11 mm long. Staminal sheath c. 12 mm long, obtuse at apex, with 2–3 mm long upper free filaments; vexillary filament c. 15 mm long; anthers dorsifixed. Ovary c. 6 mm long, hairy, with c. 2.5 mm long stipe; style slightly incurved towards apex, hairy almost throughout; stigma capitate, glabrous. Pods 12–17 x 6–7 mm, subsessile, oblong, straight, turgid, abruptly pointed at apex with 2–3 mm long beak, hairy, dehiscent along dorsal suture, partially bilobed, stipe c. 1 mm long. Seeds 3–4 x 2–3 mm, oblong to orbicular-reniform, brown, turgid, glabrous.

Flowering & Fruiting : June–August.

Habitat: On open or shady mountain slopes in dry as well as in moist sandy soils. Altitudes 2600–4000 m.

Distribution: INDIA: Himachal Pradesh, Jammu & Kashmir, Uttarakhand; PAKISTAN.

Specimens examined: **Himachal Pradesh:** Baspa valley, Ranikhanda, 3800 m, 25.5.1972, *Janardhanan* 47473 (BSD); Pangi, Chakot forest, 2743–3048 m, 16.8.1897, *Lace* 1634 (CAL); Pangi, *Dovie* 25 (CAL); Lahul, Bhaga valley, below Kelong, 3353 m, 4.8.1901, *Holland* (CAL); Chamba, August 1880, *Ellis* 382 A (CAL). **Jammu & Kashmir**: Kashmir; Nagmarg, 2743 m, 26.6.1913, *F. Koebel* 16 (CAL). **Uttarakhand**: Uttarkashi Dist.: Gangotri, 3353–3657 m, October 1881, *Duthie* 1491 (CAL, BM); Gangotri, 3300 m, 5.10.1967, *Naithani* 37498 (BSD); Gangotri, 2509 m, N31°02.288′, E79°45.580′, 3.9.2006, *Chaudhary et al.* 229557 (LWG); Bhojwasa to Gangotri, 31.7.1987, *Uniyal* & *Balodi* 80542 (BSD); Gangotri, 28.6.1986, *Raghuvanshi* 2 (LWG); Gangotri to Chirbasa, 1.6.1992, *Datt* & *Brijlal* 212240 (LWG); Chirbasa, 1.9.2000, *Kumar* 219365 (LWG); Gangotri, 3100 m, 30.8.83, *Bhattacharyya* 74668 (BSD); Chirbasa, 3500 m, 5.9.1983, *Bhattacharyya* 74945 (BSD); Chirbasa-Gangotri, 3000 m, *Rau* 51649 (BSD); Gangotri, 4 km towards Chirbasa, 3360 m, 5.9. 2002, *Khan* 223079, 223080, 223082 (LWG); Chirbasa, 3500 m, 5.9.2002, *Khan* 223087, 223092, 223093, 223095, 223096 (LWG); 2 km away from Chirbasa towards Bhojbasa, 5.9.2002, *Khan* 223091 (LWG); Between Gangotri to Chirbasa, 3360 m, 5.9.2002, *Khan* 223077, 223083 (LWG); Between Chirbasa to Bhojbasa, 3343 m, N30°58.389′, E79°02.063′, 4.9.2006, *Chaudhary et al.* 229545 (LWG); Gangotri to Chirbasa, 3313 m, N30°58.949′, E79°01.355′, 4.9.2006, *Chaudhary et al.* 229551, 229552 (LWG); 3343 m, N30°58.827′, E79°01.855′, 4.9.2006, *Chaudhary et al.* 229547 (LWG); Between Chirbasa to Bhojwasa, 5.9.2002, 3660 m, *Khan* 223090 (LWG); Bhojbasa, 3871 m, N30°57.508′, E79°03.032′, 5.9.2006, *Chaudhary et al.* 229560, 229561, 229562, 229563 229564, 229565 (LWG); Bhojwasa-Gaumukh, 3750 m, *Bhattacharyya* 74774 (BSD); Gaumukh, 3922 m, 22.9.1964, *Naithani* 37407 (BSD); Gaumukh, 3000 m, 3.6.1980, *Aswal* 6489 (CDRI); Tehri-Garhwal, Nila valley, 3048 m, 19.6 1883, *Duthie* 991 (CAL- student herb); Ganges valley, near Jungla, 2438–2743 m, 16.7.1883, *Duthie* 991a (BM).

Notes: Podlech & Zarre (2013) have treated this species as a synonym under *A. rhizanthus* Royle ex Benth., however, Anand *et al.* (2010, 2012) have clearly shown that these two species distinctly differ from each other in morphological as well as molecular markers.

5. ***Astragalus pindreensis*** (Benth. ex Baker) Ali, Kew Bull. 1958: 312. 1958; Biologia 7: 59. 1961 & in Nasir & Ali, in Fl. W. Pakistan 100: 171. 1977; Sanjappa, Legum. India 92. 1992; Podlech in Rech. f., Fl. Iranica 174: tab. 118. 1999; Kumar & Sane, Legum. Soth Asia: Checkl. 237. 2003; Anand *et al.*, Taiwania 55 (3): 198, f. 1D. 2010; Podlech & Zarre, Tax. Rev. gen. *Astragalus* (Legum.) Old World 1: 601. 2013.

A. candolleanus Royle ex Benth. var. *pindreensis* Benth. ex Baker in Hook. f., Fl. Brit. India 2:133. 1876.

A. rhizanthus var. *pindreensis* (Benth. ex Baker) Podlech in Mitt. Bot. Staatss. München 25: 183. 1988. –Type: India, Garhwal, *Falconer s. n.* (lecto K!: photo LWG!).

Plate 45

Plants perennial herbs, procumbent, densely tufted from base, stemless or with very short stems; stems 1–3 cm long, densely covered with persistent stipules and leaf rachis with quite condensed internodes, minutely hairy or glabrous at maturity. Stipules 5–7 × 2–3 mm, persistent, submembranous, adanate to petiole, ovate with acute to subacuminate tip, glabrous, ciliate towards apex, prominantely longitudinally multiveined. Leaves 2–4 cm long, imparipinnately compound; petiole 0–15 mm long, sparsely pilose; rachis not herbaceous as well as not indurate (spiny), persistent, reddish in lower half, almost terete, pilose with white spreading hairs in young condition, without leaflets and hairs with age; leaflets (5–) 8–17 pairs, 2–5 × 2–3 mm., alternate or subopposite and sparse in lower portion, opposite and very close in upper portion, subsessile, ovate, oblong, oblong- elliptic or obovate, cuneate or obtuse at base, acute, obtuse or retuse and minutely appiculate at apex, glabrous above, pilose below either uniformly only along midrib and margins, decidous. Inflorescence axillary but crowded at the tip of stems, condensed raceme, (1–) 3-flowered ; peduncle up to 2 mm long, almost flat, hairs spreading, white, only along margins or glabrescent. Bracts 4–8 mm long, persistent, submembranous, linear-lanceolate, boat-shaped with acuminate tip, glabrous or glabrescent, ciliate along margins and tip. Flowers 21–23 mm long, bright yellow; pedicle 4–5 mm. long, thinly pilose with spreading, mixed black and white hairs (prominently white). Calyx 12–15 mm. long, persistant, tubular, splitting at maturity, greenish-white with red tinge, submembranous, tube c. 9 mm long, spreading pilose with mixed white and black hairs, lobes unequal, 3.5–4 mm long, subulate-triangular on dorsal side, 5–6 mm long, linear-subulate on ventral side, all ciliate and hairy only at veins, hairs pilose, spreading and mixed black and white. Standard 21–23 mm long, backwardly deflexed, lamina 11–13 × 8.5 mm, oblong- elliptic, emarginated at apex, claw 8–12 × c. 3 mm; wing petals 20–21 mm long, lamina 9–10 × 2–2.5 mm, oblong or narrowly obovate with subobtuse tip, claw c. 11 mm. long, auricle 1–1.5 × c. 1 mm; keel petals 16–18 mm long, lamina 5–7 × c. 3 mm, oblong or obovate with obtuse tip, convex on ventral margin, claw c. 11 mm. long, auricle c. 1 mm. long. Staminal sheeth 13–15 mm. long, obtuse at apex, upper free filaments 2–3 mm. long; vexillary filament 14 mm long. Ovary 4–7 mm long, linear, densly white hairy, with c. 1 mm long stipe; style 9–10 mm long, abruptly incurved in upper portion, thinly hairy in lower portion; stigma glabrous. Pods 12–15 × 6 mm, sessile, oblong-ellipsoid, narrowed towards tip into a 3–4 mm long beak, with turgid, grooved dorsally, partially reddish, bilocular, c. 15-seeded, pilose with white, half spreading hairs. Seeds c. 3.5 × 2 m, oblong-reniform with both tips rounded, turgid, brown.

Flowering & Fruiting: June–September.

Habitat: On open hill slopes and in rocky boulders. Common in Malari (Chamoli) area.

Distribution: INDIA: Himachal Pradesh, Jammu & Kashmir, Uttarakhand; AFGHANISTAN; CHINA; NEPAL; PAKISTAN.

Specimens examined: **Himachal Pradesh**: Baspa valley, Chitkal, 3810 m, 17.7.1939, *Sherriff* 7435 (CAL); Lahul, date?, collector? *s. n.* acc. no. 114073 (CAL); Chamba, Satrundi, 3475 m, 17.6.1964, *Nair* 32422 (BSD); Spiti, Pin valley, National Park Tariya, 4000 m, 25.7.2003, *Chandra Shekhar* 103695 (BSD). **Jammu & Kashmir**: Karakoram, Zangia Harar, Hunza valley, 4267 m, 3.07.1939, *Russell* 1014 (BM); Sangla, 2591 m, 25.9.1964, *Brandis* 3799 p.p. (left side specimen) (CAL). **Uttarakhand**: Almora, Phurkiya to Pindari, 17.6.1996, *Basu* 91772 (BSD); Chamoli, Malari, 22.8.1989, *Hajra*, 80594, 87217 (BSD); Badrinath near Vasundhara fall, 3500 m, 24.8.2007, *Nayaka et al.* 225088 (LWG).

6. ***Astragalus rhizanthus*** Royle ex Benth. in Royle, Illustr. Bot. Himal. Mount. 200. 1835; Baker in Hook.f., Fl. Brit. India 2:131. 1876; Ali, Biologia 7: 57. 1961 & in Nasir & Ali, Fl. W. Pakistan 100: 170. 1977; Podlech, Mitt. Bot. Staatss. Munchen 25: 177. 1988 p. p.; Sanjappa, Legum. India 93. 1992; Kumar & Sane, Legum. South Asia: Checkl. 239. 2003; Anand *et al.*, Taiwania 55 (1): 198, f. 1A. 2010 & Proc. Indian natn. Sci. Acad. 78 (2): 174, f. 1 & 2. 2012; Podlech & Zarre, Tax. Rev. gen. *Astragalus* (Legum.) Old World 1: 620. 2013 p. p. –Type: India, Leeo in Kunawur, *Royle s. n.* (lecto LIV *n. v.*).

A. anomalus Bunge, Mém. Acad. Imp. Sci. Saint Pétersbourg 11 (16): 35. 1868 & 15 (1): 62. 1869. –Type: *Falconer s. n.* (sub nom. *A. malacophyllus*) (holo K *n. v.*).

A. dscheratuensis Širj. & Rech.f. var. *viridis* Širj. & Rech.f., Biol. Skr. **9**(3): 71. 1958. –Type: Nuristan, Netshingel, 3500m, 15.7.1949, *Edelberg* 1983 (lecto C *n. v.*; iso W *n. v.*).

A. nuristanicus Širj. & Rech. f., Biol. Skr. 9, 3: 77. 1957, non Kitam. 1956. –Type: Afghanistan, Nuristan, Parun, 31.7.1949, *Edelberg* 2017 (lecto C *n. v.*; iso W *n. v.*).

A. nuristanicus var. elasoonensis Širj. & Rech. f., Biol. Skr. 9, 3: 79. 1957. –Type: Afghanistan, Elasoon, 7.7.1949, *Edelberg* 1746 (lecto W *n. v.*; iso C *n. v.*).

A. badrinathensis Sharam *et al.*, J. Econ. Tax. Bot. 14 (1): 113. 1990. –Type: India, Uttarankhand, Chamoli dist., Vasudhara near Badrinath, 3400 m, 17.8.1986, *Sharma* 15734 (holo CDRI!; iso CDRI!).

Plate 8

Plants perennial herbs, stemless or tufted with several stems arising from the top of rootstock, stems minute to 3–10 cm long internodes indistinct, completely covered with stipules and leafbase, white pilose, glabrous to glabrescent with age. Stipules 5–17 x 1–4 mm, persistent, adnate to the petiole at base, ovate-lanceolate, acuminate at tip, sparsely to densely pilose outside, glabrous inside, sometimes only ciliate along margins, prominently longitudinally mult-veined. Leaves (3.5–) 8.5–22 (–31) cm long, very close and crowded together particularly towards the tip of stem, imparipinnately compound; petiole (3–) 4.5–7.5 cm long, white pilose with spreading or more or less forwardly oriented hairs; rachis (5.5–) 10–16 (–23) cm long, filiform, neither persistent nor indurated, hairy as petiole; leaflets 14–22 pairs, (3–) 10–15 (–19) x 3.5–6.5 mm, quite variable, subsessile, opposite, subopposite or alternate, narrowly to broadly obovate or ovate, lanceolate, oblanceolate, oblong, oblong-elliptic or sometimes almost orbicular, cuneate at base, entire along margins, acute, obtuse-apiculate, rounded to subtruncate-mucronate or occasionally emarginate at apex, glabrous or sometimes moderately pilose above, moderately to densely white, spreading, pilose below, terminal leaflet variable, usually smaller in size than other leaflets but neither deciduous nor becoming needle like. Inflorescence aggregate at tip of stems, condensed-raceme, sessile to shortly peduncled, 4–8-flowered; peduncle up to 10 mm long, spreading or appressed white pilose. Bracts 4–8 (–15) mm long, equal to or longer than pedicel, linear to linear-lanceolate, ciliate along margins, pilose outside, glabrous inside. Bracteoles sometimes present on pedicel, 2, minute to 2 mm long. Flowers 18–28 mm long, yellow; pedicel 3–6 mm long, white pilose with spreading hairs. Calyx (10–) 12–18 mm long, persistent, tubular, splitting at maturity, pilose with white, spreading hairs outside, tube 7–10 mm long, lobes (–2) 5–8 mm long, usually shorter but sometimes ± equal to tube, subequal, shorter on dorsal side, linear. Standard 18–23 mm long, lamina 9–12 x 6–10 mm, broadly obovate or oblong-elliptic, emarginate at tip, claw 8–11 x c. 1.5 mm; wing petals 17–23 mm long, shorter than or ± equal to standard, lamina c. 8 x 4 mm, oblong, rounded at tip, upper auricle c. 1 mm long, spurs present, claw c. 9 mm long; keel petals 13.5–19 mm long, shorter than wing petals, lamina 5–5.5 x c. 3.5 mm, curved in upper part on ventral margin, upper auricle minute, spurs present, claw 8.5–9 mm long. Staminal sheath c. 11 mm long, obtuse at tip, with 1-2 mm long upper free filaments; vexillary filament c. 12 mm long; anthers dorsifixed. Ovary c. 4 mm long, linear, minutely stipitate (stipe 1.5–2 mm long), densely sericeous, c. 18-ovuled; style c. 8 mm long, incurved in upper portion, glabrous; stigma capitate, glabrous. Pods 12–20 x 6–7 mm long, subsessile, oblong, straight, turgid, grooved dorsalloy, abruptly tapering into a c. 3 mm long beak at tip, moderately to densely white pilose with spreading hairs, dehiscent along dorsal suture, bilocular. Seeds 2.5–3.2 x c. 2 mm, reniform, turgid, brown or blackish-brown, smooth, glabrous.

Flowering & Fruiting: July–August (–Sepember).

Habitat: Common on grassy hill slopes, dry stony slopes, rocky grounds, along river sides and agriculture fields, in sandy and hard soils in temperate and alpine zones, at altitudes ranging from 3000–5400 m.

Distribution: INDIA: Himachal Pradesh, Jammu & Kashmir, Uttarakhand; AFGHANISTAN; CHINA; NEPAL; PAKISTAN.

Specimens examined: **Himachal Pradesh**: Lahul-Spiti: Before Khoksar, 3814 m, 28.7.2006, *Chaudhary et al.* 229407, 229408, 229409 (LWG); Khoksar proper in potato field, 3136 m, 28.7.2006, *Chaudhary et al.* 229416 (LWG); Khoksar proper towards river side, 3136 m, 28.7.2006, *Chaudhary et al.* 229417 (LWG); Khoksar, 3134 m, 28.7.2006, *Chaudhary et al.* 229421, 229422 (LWG); Khoksar to Sissu, 3200 m, 29.7.1935, *Mohindernath* (CAL); Pagni, *Stoliczka s. n.* (CAL); Khoksar, 3200 m, 16.9.2001, *Chaudhary* 220678 (LWG); Khoksar, 3150 m, 2.8.2002, *Chaudhary & Khan* 206737 (LWG); Khoksar, 3300 m, 3.8.2003, *Chaudhary & Khan* 225004 (LWG); 4 km away from Khoksar, 3200 m, 3.8.2003, *Chaudhary* & *Khan* 225007 (LWG); Khoksar, 3200 m, 23.8.1982, *Aswal* 11059 (CDRI); Khoksar, 3200 m, 2.7.1943, *Bhattacharyya* 51729 (BSD); Khoksar-Sissu, 3400 m 26.5.1957, *Rau* 5877 (BSD); Khoksar, 3200 m, 18.7.1972, *Bhattacharyya*, 48679 (BSD); Khoksar, 3200 m, 4.8.1970, *Bhattacharyya* 40524 (BSD); Khoksar-Sissu, 3400 m, 26.6.1958, *Rau* 5877 (BSD); near Chhatru, 3 Km before from Khoksar side, 3300 m, 3.8.2002, *Chaudhary* & *Khan* 206746 (LWG); Sissu, 3200 m, 17.7.1977, *Aswal* 6790 (CDRI); Gondla, 3000 m, 15.9.2001, *Chaudhary* 220669 (LWG); Gondla, 2905 m, 30.7.2006, *Chaudhary et al.* 229463 (LWG); Gondla, 2905 m, 30.7.2006, *Chaudhary et al.*, 229464 (LWG); Gondla, 2905 m, 30.7.2006, *Chaudhary et al.* 229466 (LWG); Gondla, 2905 m, 30.7.2006, *Chaudhary et al.*, 229480 (LWG); Sissu-Gondhla, 3300 m, 25.7.1979, *Aswal* 10475 (CDRI); between Gondla & Tandi, 3100 m, 3.8.2003, *Chaudhary* & *Khan* 225014 (LWG); Darcha river bed, 3600 m, 15.8.1982, *Aswal* 10948 (CDRI); Near Trilokinath temple, 3845 m, 30.7.2006, *Chaudhary et al.*, 229476 (LWG); Trilokinath, 20.8.1985, *Aswal* 15356 (CDRI); Jhala, 12000 ft, 15.6.1941, *Bor* 9722 (DD); Chota Dara, 3475 m, 31.7.2006, *Chaudhary et al.* 229485 (LWG); Between Chatru and Chota Daraa, 3027 m, 31.7.2006, *Chaudhary et al.* 229493 (LWG); Near Chatru, 3034 m, 31.7.2006, *Chaudhary et al.* 229498 (LWG); Gramphu, 2996 m, 31.7.2006, *Chaudhary et al.* 229500 (LWG); Chotadhara, 3900 m, 21.7.1972, *Bhattacharyya* 48775 (BSD); Chotadhara, 3700 m, 3.8.2002, *Chaudhary* & *Khan* 206750 (LWG); Chotadhara, 18.7.2008, *Chaudhary et al.* 249611 (LWG); 11 Km before Kunzom towards Khoksar side, 4000 m, 6.8.2002, *Chaudhary* & *Khan* 206788 (LWG); near Patscio, 3800 m, 4.8.2003, *Chaudhary* & *Khan* 225021 (LWG); Lahul, 3200 m, 2.8.1971, *Bhattacharyya* 45004 (BSD); Spiti, Pin Valley National Park, Tariya-Pin Parvati Pass, 4300 m, 24.7.2003, *Chandra Sekhar* 103672 (BSD); Spiti, Pin Valley National Park, Chhohem, 4100 m, 13.7.2003, *Chandra Sekhar* 103676 (BSD); Spiti, Pin Valley National Park, Chhohem, 4400 m, 14.7.2003, *Chandra Sekhar* 103692 (BSD); Sangla near Thinb (Baspa valley), c. 2600 m, 16.5.1972, *Janardhanana* 46757 (BSD); Baspa valley, Nagasti hill slope above, Intelligence bureau barracks, c. 3750 m, 25.5.1972, *Janardhanan* 47449 (BSD); Baspa valley,Chitkul near yak form, 24.5.1972, *Janardhanan* 47415 (BSD); Chitkul, 3810 m, 17.7.1939, *Sherriff* 7435 (BM); Rakcham, on left bank of river near glacier (Baspa valley), c.3505 m, 20.5.1972, *Janardhanan* 47343 (BSD); Chini, 3200 m, 4.6.1962, *Nair* 22410 (BSD), Pangi, Bhut Nullah, Tau, 3353 m, 19.6.1981, *Sayers* 3731(BM). **Jammu & Kashmir**: Karakoram: Sokha Glacier, right bank, 4267–4572 m, 24.8.1939, *Russell* 1705 (BM); Kero Lugma Glacier, left bank, 3048–3353 m, 26.7.1939, *Russell* 1352 (BM); Haramosh range, Mani Basin, 12.8.1957, *Culbert* 11 (BM); Without locality, 3505 m, 14.8.1876, *Clarke* 30470 (BM); Sinel valley below Baltal, 2743–3048 m, 27.6.92. *Duthie* 11572 (BM); Sassir Pass, 4480 m, July 1925, *Meinertzhagem s. n.* (BM); Ladak, Leh, Khardung La, 4724 m, 18.6.1941, *Ludlow* & *Sherriff* 8424 (BM); Ladakh, Rungdum Gompa, 3962 m, July 1905, *Meebold* 1367 (CAL); Ladakh, Dras, 3100 m, 4.8.1980, *Bhattacharyya* 71363 (BSD); Dras, 3048 m, 7.6.1977, *Stainton* 7793 (BM); Near Dras town, 3091 m, N 34°25.843′E 75°44.693′, 5.8.2011, *Chaudhary et. al.* 263880 (LWG); Kashmir, Date?, *Nair* 37015 (BSD); Gulmarg, *Prescott-Decie s. n.* (BM); Dras, 3353 m, ft., 30–31.8. 1922. *Stewart*-separated from no. 7402 (BM); Kashmir, Pachtsani, 3600 m, 27.7.1966, *Nair* 37015 (BSD); Ladakh, Lakong, 4500 m, 23.7.1973, *Bhattacharyya* 52183 (BSD); Ladakh, Matyan, 3330 m, 17.7.1976, *Wadhwa* 58688 (BSD); Ladakh, between Leh & Khardunga La, 5000 m,

8.8.2003, *Chaudhary & Khan* 225052 (LWG); Pahlgam, 7000 ft.,1909, *Douie* 1 (CAL); Zojila, 3962 m, 31.7.1891, *Gammie* (CAL); Kashmir, 1864, *Falconer* 423 (CAL); Gilgit, 1885, *Giles s. n.* (CAL); Kungwatan, 3000 m, 9.6.1959, *Rao* 9286 (BSD); Kishtwar, Bamaj nullah, near Sapphire mines, 3353 m, 11.7.43, *Ludlow & Sheriff* 9179 (BM); Bringli valley, 2743 m, 15.5.1939, *Ludlow* 16A (BM); Kishenganga valley and the road to Naga Parbat, Rattu to Rupal Nullah, 2591 m, 22 8.1939, *R. R. & I. D. Stewart* 18853 (BM). **Uttrakhand**: Garhwal, Badrinath, 4.7.1957. *Sarin & Rau* 2959 (BSD); Tehri Garhwal, Ganges valley near Jangla, 2450–2750 m, 16.7.1883, *Duthie* 991 (K).

Notes: A notoriously variable species especially in stems, leaves and hairs. The stem is generally absent to occasionally well-developed. The length of leaves and number of leaflets vary considerably in the same population. The variations have been observed in the size and shape of leaflets even on the same plant. The density and orientation of hairs on stipules, leaf-rachis, leaflets, calyx and pods also vary remarkably from population to population.

5. *Astragalus* sect. *Cenantrum*

Bunge, Mém. Acad. Imp. Sci. Saint Pétersbourg 11 (16): 23. 1868; Podlech & Zarre, Tax. Rev. gen. *Astragalus* (Legum.) Old World 1: 170. 2013. –Lectotype species: *Astragalus frigidus* (L.) A. Gray.

Phaca L., Sp. Pl. 755. 1753. –Type species: *Phaca alpina* L.

Erect subshrubs, caulescent. Hairs basifixed. Stipules greenish, free from petiole, free from each other or occasionally basal one connate opposite to petiole. Leaves imparipinnately compound, nearly sessile or with a short petiole. Racemes long pedunculate, few to many-flowered. Bracts persistent. Bracteoles absent. Flowers yellow or pale greenish-yellow. Calyx campanulate to shortly tubular, laterally gibbous at base, obliquely cut at mouth. Wing petals obtuse or rarely emarginated at apex; keel petals shorter or sometimes longer than wing petals. Stamens diadelphous or rarely monadelphous. Style and stigma glabrous. Pods more or less stipitate, unilocular, valves papery, glabrous or hairy.

Key to the Species

1a. Leaflets very narrowly oblong, obovate or ovate, generally 4–5 times longer than width .... **1. *A. floridulus***

1b. Leaflets comparatively wider, generally 2–3 times longer than width 2

2a. Ovary glabrous; pods glabrous, compressed ....**3. *A. pseudochlorostachys***

2b. Ovary hairy; pods hairy, turgid 3

3a. Plants decumbent to erect, 6–35 cm long; peduncle distinctly longer than subtending leaf; flowers 10–11 mm long; calyx 4–5 mm long, tube glabrous, teeth 1.5-3 mm long, narrowly triangular with acuminate tip ....**2. *A. lessertioides***

3b. Plants erect, 30–60 cm long; peduncle more or less equal to subtending leaf; flowers 20–23 mm long; calyx 10–11 mm long, tube hairy, teeth up to 1.5 mm long, triangular ....**4. *A. tecti-mundi***

1. *Astragalus floridulus* Podlech, Sendtnera 1: 270. 1993; Kumar & Sane, Legum. South Asia: Checkl. 229. 2003; Podlech & Zarre, Tax. Rev. gen. *Astragalus* (Legum.) Old World 1: 181. 2013.

A. floridus Benth. ex Bunge, Mém. Acad. Imp. Sci. Saint Pétersbourg 11 (16): 24. 1868 & 15 (1): 28. 1869, non Scop. 1786; Baker in Hook. f., Fl. Brit. India 2: 127. 1876; Grierson & Long in Grierson & Long, Fl. Bhutan 3: 718. 1987; Sanjappa, Legum India 87. 1992; Press *et al.*, Anno. Checkl. Flow. Pl. Nepal 166. 2000. –Type: India, Himalaya, Sikkim, Tunga, 13000 ft., *Hooker s. n.* (lecto P *n. v.*; iso K! , P *n. v.*, W *n. v.*, CAL!: photo LWG!).

Plants erect herbs, stems elongated, 15–30 cm high, thinly pilose with black (mixed black & white in lower portion), mostly appressed (a few long hairs spreading) hairs. Stipules 3–6 × 1–2 mm, free, lanceolate, densely ciliate with long, blackish-brown, spreading hairs along margins. Leaves 5–7 mm long, imparipinnately compound; petiole minute to 10 mm long, petiole and rachis pilose with mostly black (sometimes intermixed with few white hairs), appressed hairs; leaflets 7–10 pairs, 6–12 × 1.5–3 mm, opposite, subopposite or alternate, generally narrowly oblong or sometimes narrowly ovate or obovate, obtuse or cuneate at base, acute or subacute at apex with a mucro, glabrous above, pilose with appressed, white hairs below. Inflorescence axillary, peduncled raceme, c. 11 mm long, longer than subtending leaves, raceme many-flowered, close; peduncle 6.5–8 cm long, pilose with blackish-brown, appressed hairs, more hairy towards tip. Bracts c. 4 mm long, persistent, linear, pilose with blackish-brown, long hairs along margins. Flowers c. 12 mm long, yellow, lower ones nodded; pedicel c. 5 mm long, densely blackish-brown hairy. Calyx 6–7 mm long, borne obliquely, almost pouched above attachment with pedicel, campanulate, persistent, densely pilose with long, blackish-brown (mixed with a few white hairs), appressed hairs outside, tube c. 4 mm long, lobes c. 2 mm long, lanceolate, more hairy on both sides. Standard c.12 × 4–5 mm, obovate-elliptic, emarginate at apex; wing petals c. 7 × 1 mm, oblong; keel petals c. 7 × 3 mm. Ovary stipitate, densely hairy. Pods 18–20 × 6–7 mm, ellipsoid, narroed at both ends, stipitate, densely pilose with blackish-brown, appressed hairs, unilocular, 3–5-seeded, stipe c. 4.5 mm long, equal to calyx tube or slightly exceeding the tube but shorter than calyx length.

Flowering & Fruiting: July–August.

Habitat: Altitudes 4200–4700 m.

Distribution: INDIA: Sikkim; BHUTAN; CHINA; NEPAL.

Specimens examined: **Sikkim**: Llonak, 28.7–8.1903, *Younghusband* 214 (CAL).

2. ***Astragalus lessertioides*** Benth. ex Bunge, Mém. Acad. Imp. Sci. Saint Pétersbourg 11 (16): 24. 1868 & 15 (1): 28. 1869; Baker in Hook. f., Fl. Brit. India 2: 124. 1876; Grierson & Long in Grierson & Long, Fl. Bhutan, 1 (3): 718. 1987; Sanjappa, Legum India 90. 1992; Press *et al.*, Anno. Checkl. Fl. Pl. Nepal 166. 2000; Kumar & Sane, Legum. South Asia: A Checkl. 233. 2003; L.B.Chaudhary, Rheedea 15 (1): 76. 2005; Podlech & Zarre, Tax. Rev. gen. *Astragalus* (Legum.) Old World 1: 188. 2013. –Type: India, Sikkim, in jugo himalayensis prov. Sikkim, regione alpina, 12000–14000 ft., *Hooker s. n.* (holo P *n. v.*; iso K!, L *n. v.*, P *n. v.*, W *n. v.*).

Plate 46

Plants decumbent to erect herbs, 6–35 cm long, stems 2–18 cm long, tinged with red colour, simple or once or twice branched, terminated into solitary, short to long peduncled Inflorescence, glabrous or thinly pilose with white, spreading hairs. Stipules 5–15 × 2–6 mm, tinged with red colour, foliaceous, free, oblong-lanceolate or narrowly ovate, obtuse at tip, ciliate with long, mixed white and black or only white, spreading, forwardly oriented hairs, otherwise glabrous. Leaves 2.5–8 cm long, impripinnately compound; petiole 1.0–2.5 cm long; rachis and petiole deeply channeled above, glabrous; leaflets 4–6 pairs, 6–14 (–25) × 3–8 (–10) mm, opposite, obovate, ovate, elliptic or oblong, obtuse or cuneate at base, obtuse or rounded at apex, glabrous above, unevenly thinly pilose below, ciliate with white, forwardly oriented hairs along margins and extra margins. Inflorescence axillary, peduncled capitate raceme (head), close, 7– 20-flowered; peduncle 7–10 cm (up to 18 cm in fruits) long, tinged with red colour, distinctly longer than subtending leaf, glabrous in lower portion, pilose with blackish-brown, appressed hairs in upper portion. Bracts 3–7 × 0.5–1 mm, persistent, distinctly longer than pedicel, linear or linear-lanceolate, densely pilose with long, forwardly oriented, mixed brown and white or black hairs outside and along margins. Flowers 10–11 mm long, dark yellow, nodded at maturity; pedicel less than 2 mm long, densely pubescent with blackish-brown, forwardly oriented hairs. Calyx 4–5 mm (sometimes 8–9 mm in fruits in *Duthie* 984) long, campanulate, persistent, tube 2.5–3 mm long, glabrous, lobes c. 1.5 mm long, pubescent with blackish-brown hairs on both sides (more densely hairy inside). Standard 11 × 7.5 mm, obovate, emrginate at apex, claw not distinct; wing petals equal to standard, lamina c. 4 × 1.5 mm, oblong, obtuse at apex, upper auricle minute, claw c. 7 mm long; keel petals almost equal to standard, lamina c. 4 × 2 mm, oblong, slightly curved towards apex, subacute at apex, upper auricle minute dentate along margin, claw c. 7 mm long. Staminal sheath c. 10 mm long, obtuse at apex, free filaments c. 1 mm long, alternately longer and shorter; vexillary filament c. 10 mm long;. Ovary stipitate, densely hairy with mixed white and black or only brownish hairs, with c. 3 mm long stipe. Pods 12–15 (–20) × 4– 6 mm, lanceolate,narrowed at both ends, turgid, stipitate, unilocular, densely pilose with blackish-brown, forwardly oriented hairs, 3–4-seeded, stipe 2.5–3 mm long, as long as calyx tube or slightly longer than tube but shorter than calyx.

Flowering & Fruiting: July–September (–October).

Habitat: On grassy hillsides or on flat lands in moist soil at altitudes ranging from 3000–5000 m.

Distribution: INDIA: Sikkim, Uttarakhand; BHUTAN; CHINA; NEPAL.

Specimens examined: **Sikkim**: North Sikkim, Thangu, Giagaon, 4600 m, 13.8.2004, *Chaudhary* 225072 (LWG); Giagaon, 4293 m, N°27 56.174′ E88° 33.718′, 4.9. 2015, *Chaudhary & Kushwaha* 259386 (LWG); (LWG);Llonak, 3962 m, 26.7.1909, *Smith & Cave* 1773 (CAL); Llonak, 3962 m, 26.7.1909, *Smith & Cave* 1173 (LBG); Without locality, 1892, *Gammie* p.p. (CAL); Lachung, Samdong, 5029 m, 17.8.1892, *Gammie* 852 (CAL); Tebli, 4572 m, *Lepcha collector* 2722 (CAL, LBG); Zemu valley, 3657 m, 12.7.1909, *Smith & Cave* 1224 (CAL, LBG); Llonak, Naka Chu, 4877 m, 3.8.1909, *Smith & Cave* 1896 (CAL). **Uttarakhand**: Tehri-Garhwal, Chinpul, opposite Banderpunch, 3657–3962 m, ft., *Duthie* 984 (CAL, BM); Kumaon, Furkia, Pindari moraine, 3500–5000 m, 23.10.1957, *Rao s. n.* (BSD).

3. ***Astragalus pseudochlorostachys*** Ali, Bot. Notis. 118: 94. 1965 & in Nasir & Ali, Fl. W. Pak.istan 100: 174. 1977; Sanjappa, Legum. India 93. 1992; Podlech in Rech. f., Fl. Iranica 174: t. 39. 1999; Kumar & Sane, Legum. South Asia: Checkl. 237. 2003; Podlech & Zarre, Tax. Rev. gen. *Astragalus* (Legum.) Old World 1: 201. 2013. –Type: Pakistan, Khotan, N of Mastuj, between Turikho and Mastuj rivers, 11000 ft., 11.7.1958, *Stainton* 2861 (holo BM!; iso G *n. v.*, LE *n. v.*, RAW *n. v.*).

Plate 47

Plants herbs, erect, 25–40 cm high, stems puberulous with completely appressed, forwardly oriented, white hairs below, glabrescent to glabrous above. Stipules 8–10 × 3–5 mm, foliaceous, free, almost completely encircling the stem, glabrous, sometimes margins ciliate with appressed black hairs. Leaves 6–9 cm long, imparipinnately compound; petiole 6–15 mm long; rachis 3–7 cm long, channeled above, glabrous; leaflets 4–5 pairs, 20–36 × c. 11 mm, subsessile, oblong, oblong-elliptic, narrowly ovate or ovate, acute at apex, glabrous on both sides or glabrous above, pubescent with white, appressed hairs only along margins and midrib or throughout below. Inflorescence axillary, peduncled raceme, longer than subtending leaf; raceme 6–7-flowered, close, peduncle 8–10 cm long, glabrescent to glabrous below, pubescent with half appressed, blackish-brown hairs towards tip. Bracts 6.5–9 × 2–6 mm, persistent, foliaceous, ovate or ovate-lanceolate, longer than pedicel, pilose outside with blackish-brown hairs, ciliate along margins. Flowers c. 20 mm long, pale yellow; pedicel 2.5–4 mm long, pilose with half appressed, black hairs. Calyx c. 9 mm long, tubular, oblique at mouth, tube c. 8 mm long, thinly pubescent with black, appressed hairs outside, glabrous inside, lobes 1.5–2 mm long, deltoid, densely hairy with black hairs on both sides. Standard 18–20 × c. 10 mm, broadly elliptic with narrow base and faintly retuse apex; wing petals c. 16 mm long, shorter than standard, lamina c. 7 × 2 mm, narrowly ovate with obtuse tip, upper auricles very short, spurs present, claw c. 9 mm long; keel petals c. 14 mm long, shorter than wing petals, lamina c. 5.5 × 3 mm, slightly curved towards apex, upper auricle absent, spurs present claw c. 8.5 mm long. Staminal sheath c. 12 mm long, free filaments 1.5–3 mm long, shorter and longer alternately arranged; vexillary filament c. 12.5 mm long. Ovary c. 6 mm long, linear, glabrous, c. 11-ovuled, stipitate, with c. 6 mm long stipe; style c. 1.5 mm long, abruptly incurved from base, glabrous; stigma capitate. Pods c. 17 × 5 mm, oblong-ovate, compressed, glabrous, 4–5-seeded, stipe c. 7 mm long, slightly shorter than calyx.

Flowering & Fruiting: July–August.

Distribution: INDIA: Jammu & Kashmir; PAKISTAN.

Specimens examined: **Jammu & Kashmir**: Ladakh, Matyan, 3330–3570 m, 17.7.1976, *Wadhwa* 58650 (BSD).

Notes: Podlech & Zarre (2013) have not recorded it from India. They have reported it only from N. Pakistan (including Baltistan and Gilgit).

4. *Astragalus tecti-mundi* Freyn, Bull. Herb. Boiss., sér. 2, 4: 456. 1904; Podlech in Rech. f., Fl. Iranica 174: t. 38. 1999; Kumar & Sane, Legum. South Asia: Checkl. 242. 2003; Podlech & Zarre, Tax. Rev. gen. *Astragalus* (Legum.) Old World 1: 205. 2013. –Type: Tadzhikistan, Pamir, Kisil Krashim ad fl. Pamir-Daria, 3600 m, *Paulsen* 1260 (lecto C *n. v.*; iso BRNM *n. v.*).

A. frigidus auct non (L.) A. Gray 1864; Baker in Hook. f., Fl. Brit. India 2: 130. 1876; Ali, Biologia 7: 38. 1961& in Nasir & Ali, Fl. W. Pakistan 100: 178. 1977. Sanjappa, Legum. India 87. 1992; Press *et al.*, Annot. Checkl. Fl. Pl. Nepal 167. 2000.

Plants erect herbs, 30–60 cm long, stems glabrous to glabrescent, with long internodes. Stipules 9–17 × 2–6 mm, free, connate before petiole in basal portion of plant, persistent, spreading, membranous, oblong, lanceolate or ovate, acute at tip, sparsely pilose with white hairs outside, glabrous inside or glabrous except margins and tip. Leaves 10.5–14 cm long, imparipinnately compound; petiole 2–35 mm long; rachis 5–8.5 cm long, glabrous to glabrescent; leaflets 4–7 pairs, 21–37 × 6–11 mm, opposite, sparse, subsessile, ovate, oblong or oblong elliptic, acute or subobtuse to obtuse with a very minute mucro at tip, glabrous on both sides or glabrous above, sparsely pubescent with white hairs below, sometimes intermixed with black hairs Inflorescence 13–17 cm long, equal to or longer than subtending leaves, axillary, long peduncled raceme, many-flowered; peduncle 9.5–12 cm long, glabrescent with white hairs below, glabrescent to pubescent with black hairs towards tip. Bracts 7–10 × 2 mm, longer than pedicel, persistent, deflexed, similar to stipules, glabrescent outside, glabrous inside, margins and tip hairy with spreading black hairs. Flowers 20–23 mm long, yellow, nodded; pedicel 3–5 mm long, densely pubescent with black, half appressed hairs. Calyx 10–11 mm long, persistent, tubular, tube 9–10 mm long, sparsely to moderately pilose with black or brown spreading hairs outside, glabrous inside, lobes minute to 1.5 mm long, densely villous on both sides with black hairs. Standard c. 23 × 10 mm, deflexed, spatulate, emarginate at apex; wing petal c. 21 mm long, shorter than standard, lamina c. 8 × 3 mm, oblong with obtuse tip, slightly backwardly curved, upper auricle c. 2 × 1 mm, triangular, claw c. 13 mm long; keel petals c. 19 mm long, shorter than wing petals, c. 6 × 3 mm, slightly curved towards apex, upper auricle minute, claw c. 13 mm long. Staminal sheath c. 17 mm long, with truncated apex, free filaments, 1.5–2.5 mm long, alternately shorter and longer; vexillary filament c. 18 mm long. Ovary c. 7 mm long, densely hairy, with c.10 mm long stipe; style c. 3 mm long; stigma capitates, glabrous. Pods 23–30 × c. 8 mm, oblong-ellipsoid, tapering at both ends, turgid, unilocular, pubescent with black appressed hairs, stipe 9–10 mm long, equal to calyx. Seeds c. 3 × 2 mm, reniform, turgid, smooth, blackish-brown, glabrous.

Notes: It is a widely distributed species and shows enormous amount of variations. The stipules are quite variable even on the same plant. They are connate and almost encircling the stem at few nodes in the basal portion of the plant, while others are completely free. The stipules found in upper portion of plants are narrower (lesser wider) than the stipules found in lower portion. They also vary in shape, size, apex and hairiness from population to population. Baker (1876) describes pods glabrous while in the present study pubescent pods have been observed like Ali (1977).

Podlech & Zarre (2013) have recognized two subspecies i. e. subsp. *tecti–mundi* and subsp. *orientalis* Podlech based on following characters. All specimens observed in the present study belong to subsp. *orientalis*. I could not observe specimens of subsp. *tecti-mundi*.

4a. subsp. *tecti-mundi*

Leaflets 4–5 pairs, glabrous on both sides. Standard 14–17 mm long. Pods 10–25 mm long, flat on both margins.

Flowering & Fruiting: July–September.

Distribution: INDIA: Jammu & Kashmir; NE AFGHANISTAN; NEPAL; PAKISTAN; TADZHIKISTAN; W. CHINA.

Specimens examined: Nil.

Notes: Podlech & Zarre (2013) have recorded it from Jammu & Kashmir (Zanskar-Umasi Lah and N. Sanmodangra) based on the specimens housed at BM and MSB.

4b. subsp. *orientalis* Podlech, Sendtnera 7: 178. 2001; Kumar & Sane, Legum. South Asia: Checkl. 242. 2003; L.B.Chaudhary & S.K.Srivast., Taiwania 52 (1): 30, f. 8. 2007; Podlech & Zarre, Tax. Rev. gen. *Astragalus* (Legum.) Old World 1: 206. 2013.. –Type: Pakistan, Hunza- und Nagar-Gebiet, Daintar, 3300–3900 m, 1959, *Lobbichler* 115 (holo M *n. v.*; iso MSB *n. v.*).

Plates 9, 48

Leaflets 5–7 pairs, glabrous on upper surface, sparsely to loosely appressed to subappressed pilose white hairy on lower surface. Standard 15–25 mm long. Pods 25–45 mm long, slightly keeled ventrally, slightly grooved dorsally.

Flowering & Fruiting: July–September.

Habitat: On moist hill slopes in alluvial soil among grasses in open. Sometimes dominant near *Salix* plantation. Altitudes 3200–4400 m.

Distribution: INDIA: Himachal Pradesh, Jammu & Kashmir, Uttrakhand; PAKISTAN; S. CHINA.

Specimens examined: **Himachal Pradesh**: Lahaul, Kardang, *Jaeschke* (CAL); Lahul, *Jaeschke* 265 (CAL); Lahul, Between Baralacha to Sarchu, 5.8.2014, *Chaudhary et al. s. n.* (LWG); Pangi, 11000–16000 ft., 1879, *Heyde s. n.* (CAL); Pangi, *Dovie* 20 (CAL). **Jammu & Kashmir**: Karakoram: Makerum, Hispar glacier, 3810–4115 m, 2.8.1939, *Russell* 1419 (BM); Kero Lunga valley, left bank, Oblation valley, 3657–3962 m, 24.7.1939, *Russell* 1313 (BM); Sokha Glacier, left bank, 3962 m, 28.8.1939, *Russell* 1671 (BM); Hispar glacier, left bank, Turmun-Makerum, 3353–3505 m, 18.7.1939, *Russell* 1234 (BM); Ladakh, Parkhachik to Rangdom, 1.8.1988, *Chowdhery & Uniyal* 85980 (BSD); Bhotkalla, Cashem, 3505 m, July 1905, *Meebold* 1383 (CAL); Kishtwar, Barnaj Nullah, Sapplier Mines, 3048 m, 9.9.1943, *Ludlow & Sherriff* 9162 (BM).

Notes: In the present study no specimens have been observed from Uttarakhand, however, Podlech & Zarre (2013) have recorded it from Uttarakhand based on the specimens available at BM (Garhwal, Suhki across Basmuru and Chaia Pass to Kharsali and Kumaon, Byans, Palang Gadh).

6. *Astragalus* sect. *Chaetodon*

Bunge, Mém. Acad. Imp. Sci. Saint Pétersbourg 11 (16): 136. 1868; Podlech & Zarre, Tax. Rev. gen. *Astragalus* (Legum.) Old World 3: 2012. 2013. –Type species: *Astragalus chaetodon* Bunge.

Perennial herbs, subacaulescent or with well developed stems. Hairs medifixed. Stipules adnate to petiole, free from each other. Leaves imparipinnately compound, nearly sessile or with a short petiole. Racemes subsessile to long pedunculate. Bracts persistent. Bracteoles absent. Calyx tubular, obliquely gibbous at base and narrowed towards mouth, inflated, ruptured by pod or not. Standard obovate, constricted below the middle; wing petals as long as standard. Pods sessile or subsessile, often enclosed in calyx, subbilocular, hairy.

1. ***Astragalus breviscapus*** B. Fedtsch.,Trudy Imp. St.- Petersburgsk 24: 234. 1905; Ali, Biologia 7: 69. 1961& in Nasir & Ali, Fl. W. Pakistan 100: 199, f. 26 E. 1977; Gontsch. & Popov in Komarov, Fl. U.R.S.S. 12: 822. 1946; Sanjappa, Legum. India 85. 1992; L.B.Chaudhary & S.K.Srivast., Taiwania 52: 26. 2007; Podlech & Zarre, Tax. Rev. gen. *Astragalus* (Legum.) Old World 3: 2016. 2013. –Type: [Kirgizstan] Khr. [Mountain] Managildy bei Alabugi, 6000–7000 ft., 4.6.1880, *Regel s. n.* (holo LE *n. v.*).

A. oophorus Freyn, Bull. Herb. Boiss. Ser. 2, 5: 1019. 1905, non S. Wats 1871. –Type: Tadzhikistan, Pamir, Jashil-Kul, 3800 m, 4.8.1898, *Paulsen* 1026 (lecto BRNM *n. v.*; iso C *n. v.*, LE *n. v.*).

Plants Perennial shrubs, several stems arising from the top of woody rootstock, stems reduced, c. 3 cm long, densely sericeous with white, appressed as well as spreading, medifixed hairs. Stipules free from petiole, lanceolate-triangular, pilose outside and along margins with long, white, spreading, medifixed hairs. Leaves 5.5–7 cm long, imparipinnately compound; petiole c. 3 cm long, hairy as rachis; rachis 5–5.5 cm long, pilose with long, white, spreading, medifixed hairs; leaflets 4–5 pairs, 7–15 × 3–6 mm, subsessile, subopposite to opposite, elliptic or obovate, cuneate at base, acute at apex, densely hairy below, moderately hairy above, hairs white, long, spreading, medifixed. Inflorescence axillary, peduncled raceme, raceme many-flowered, 3–9 cm long, very dense, oblong; peduncle 5–25 mm long. Bracts c. 9 × 1 mm, linear, densely covered with mixed black and white, spreading, medifixed (with distinctly unequal arms) hairs. Flowers c. 20 mm long, pinkish-violet, subsessile. Calyx c.15 mm long, tubular, becoming ovoid-inflated and membranous in fruiting, tube c.11 mm long, densely covered with white, long, spreading, medifixed hairs, lobes 4–5 mm long, subulate, covered with mixed black and white hairs. Standard c. 22 mm long, lamina c. 15 × 5 mm, obovate-stipitate, claw c. 7 mm long; wing petals c. 21 mm long, lamina c. 5 × 2 mm, oblong with subacute apex, upper auricle minute, claw c. 15 mm long; keel petals c. 19 mm long, lamina c. 4 × 3 mm, oblong with obtuse apex, incurved towards apex, claw c. 15 mm long. Staminal sheath c.16 mm long, free filaments c. 3 mm long; vexillary filament c. 18 mm. Ovary c. 7 mm, with c. 1 mm long stipe; style c. 14 mm long; stigma glabrous. Pods 8–12 × 3.5–4 mm, subsessile, narrowly oblong with a c. 3 mm long beak, compressed laterally, deeply grooved dorsally, subbilocular, covered with white, long hairs.

Flowering & Fruiting: April–July.

Distribution: INDIA (Jammu & Kashmir-Gilgit); CHINA; KIRGISTAN; PAKISTAN; TADZHIKISTAN.

Specimens examined: **Jammu & Kashmir**: Baltistan, Holshal near Nagar, Pentes seches dirigees au nord, 3100–3250 m, 20.7.1954, *Schmid* 1959 (BM).

Notes: It has been reported so far only from Gilgit region of Jammu & Kashmir with a very few collections. In the present study no specimens have been observed in the Indian herbaria.

7. *Astragalus* sect. *Coluteocarpus*

Boiss., Fl. Or. 2: 11. 1872; Podlech & Zarre, Tax. Rev. gen. *Astragalus* (Legum.) Old World 1: 245. 2013. –Type species: *Astragalus coluteocarpus* Boiss.

Plants erect, stems well developed and stout. Stipules free, but may be widened at base. Racemes long peduncled, many-flowered. Bracts and bracteoles present. Stigma ciliate. Pods distinctly stipitate, inflated, glabrous, partially to fully bilocular.

1. ***Astragalus coluteocarpus*** Boiss., Diagn. pl. or. ser.1, 9: 65. 1849 & Fl. Or. 2: 27. 1872; Gontsch. in Komarov, Fl. U. S. S. R. 12: 32. 1946; Ali, Kew Bull. 1958: 304. 1958; Biologia 7: 13. 1961; Bot. Notiser 118: 87. 1965 & in Nasir & Ali, Fl. W. Pakistana 100: 130. 1977; Wenninger, Mitt. Bot. Staats. Munchen 3o: 46, f. 4 b-d, 16 a-b, e-f. 1991; Sanjappa, Legum. India 86. 1992; Kumar & Sane, Legum. South Asia: Checkl. 225. 2003; L.B.Chaudhary & S.K.Srivast., Taiwania 52 (1): 27. 2007; Podlech & Zarre, Tax. Rev. gen. *Astragalus* (Legum.) Old World 1: 246. 2013.

Plants perennial tufted large herbs or small shrubs, erect, up to 100 cm high, stems terete, striate, glabrous. Stipules 5–7 × 2–3 mm, lateral, free, lanceolate, faintly veined, membranous, glabrous, ciliate along margins. Leaves 11–18 cm long, imparipinnately compound, petiole 15–18 mm long; rachis 10.5–15 cm long, channeled above; petiole & rachis glabrous at maturity, pubescent in young condition; leaflets 5–9 pairs, 10–14 × 6–10 mm, subopposite to opposite, obovate, elliptic, broadly-oblong or rarely ovate, cuneate at base, entire along margins, obtuse, truncate-mucronate or rarely acute at apex, glabrous on upper surface, sparsely hairy on lower surface or hairy on both sides, midrib sparsely pilose with white, almost appressed, forwardly oriented hairs; petiolule less than 1 mm long, pubescent. Inflorescence 17– 20 cm long, axillary, long peduncled raceme, many-flowered; peduncle 8–9 cm long, glabrous to very sparsely pilose with white, thin, minute, appressed, forwardly oriented hairs; floral rachis pilose with mixed white and blackish-brown or black, appressed hairs, with irregularly arranged flowers (sometimes 2–3 flowers arising from single node). Bracts c. 7 × 1 mm, distinctly longer than pedicel, caducous, linear-lanceolate, sparsely pilose outside with brown, appressed hairs (white hairs on tip). Bracteoles 2, 3–6 mm long, situated on the top of the pedicel just below the calyx, linear, pilose, hairs brown, spreading. Flowers c. 15 mm long, pale yellow; pedicel c. 2.5 mm long, pubescent with black, appressed hairs. Calyx c. 10 mm long, persitent, tubular, tube c. 5 mm long, irregularly sparsely pilose outside with black hairs, lobes unequal, upper ones c. 2.5 mm long, lower ones 4–5 mm long, all pilose outside with spreading hairs. Standard c. 15.5 mm long, deflexed, lamina c. 11.5 × 7 mm, obovate, emarginate, claw c. 4 mm long; wing petals c. 16 mm long, more or less equal to standard, lamina c. 9 × 2 mm, narrowly ovate, subobtuse, upper auricle oblong with obtuse tip, claw c. 7 mm long; keel petals c. 13.5 mm long, much shorter than standard and wing petals, lamina c. 6.5 × 5 mm, abruptly incurved, upper auricle minute, spurs faintly developed, claw c. 7 mm long. Staminal sheath c. 11 mm long, obtuse at tip, free filaments 3–4.5 mm long; vexillary filament c. 13 mm long. Ovary c. 7 mm long, with 6 mm long stipe, linear, glabrous; style c. 4 mm long, abruptly incurved, glabrous; stigma capitate, glabrous. Pods c. 27–40 × 9–10 mm, stipitate, oblong, turgid, pointed at both ends, partially septate, glabrous, faintly veined, dehiscent, 8–16-seeded, stipe 8–13 mm long. Seeds c. 4 × 3 mm, reniform, turgid, black, spotted, gabrous.

Notes: Ali (1958) has described a variety *A. coluteocarpus* var. *glaber* Ali based on pubescence of leaflets and reported it form India (Jammu & Kashmir-Dras and Himachal Pradesh-Spiti), Pakistan (Chitral), Rassia and Turkestan . In var. *glaber* the leaflets are glabrous above (adaxial) and sparsely pilose below (abaxial) while in var. *coluteocarpus* they are pubescent on both sides. Baker (1876) has mentioned only glabrous leaflets in *A. chitralensis.* Wenninger (1991) has treated var. *glaber* at subspecies level (i. e. subsp. *chitralensis* Wenninger) and recorded it only from Pakistan, as also done by Podlech & Zarre (2013).

1a. subsp. *coluteocarpus*. –Type: Afghanistan, in regno Cabulico, *Griffith* 1064 (holo K: hb Hooker *n. v.;* iso K *n. v.*, CAL!).

Plate 10

Leaflets hairy on both sides or glabrescent to hairy above, hairy below.

Flowering & Fruiting: June–July.

Distribution: INDIA: Himachal Pradesh, Jammu & Kashmir; AFGHANISTAN; PAKISTAN; TADZHIKISTAN.

Specimens examined: **Himachal Pradesh**: Upper Kunawar, 3048 m, August 1852, *Thomson* s.n. (K); Kunawar, July 1844, *Munro s. n.* (K); Kunawar, 1830–2438 m, *Thomson s.n.* (K). **Jammu & Kashmir**: Leh-Ladakh, Near Dras town, 3091 m, N 34°25.843′ E 75°44.693′, 5.8.2011, *Chaudhary et al.* 263881 (LWG); Ladakh, Western Tibet, Dras, 1.9.1922, 11000 ft., *Stewart* 7473 (K); Suru valley, Near Pharma village, 2903 m, N 34°22.378′ E 75°58.323′, 4.8.2011, *Chaudhary et al.* 263856 (LWG).

Notes: Ali (1961, 1977) and Wenninger (1991) have also reported it from Jammu & Kashmir (Dras, Astor) and Himachal Pradesh (upper Kunawar). However, Podlech & Zarre (2013) have not recorded it from India.

1b. subsp. *chitralensis* Wenninger, Mitt. Bot. Staatss. Munchen 30: 52. 1991; Kumar & Sane, Legum. South Asia: Checkl. 225. 2003; Podlech & Zarre, Tax. Rev. gen. *Astragalus* (Legum.) Old World 1: 247. 2013.

A. coluteocarpus var. *glaber* Ali, Kew Bull. 1958: 304. 1958; Biologia 7: 14. 1961& in Nasir & Ali, Fl. W. Pakistana 100: 130. 1977; Sanjappa, Legum. India 86. 1992. –Type: Pakistan, Chitral, Barum Gol, below Zapotili, 11480 ft., 23.6.1950, *Wendelbo s. n.* (holo K!; iso W *n. v.*).

Plates 49, 50

Leaflets glabrous on upper side, sparsely hairy on lower side.

Flowering & Fruiting: June–September.

Habitat: On exposed sandy or rocky soils. Altitudes: 3000–3800 m.

Distribution: INDIA: Himachal Pradesh, Jammu & Kashmir, Uttarakhand; PAKISTAN; NORTHERN EURASIA.

Specimens examined: **Himachal Pradesh:** Spiti, Mani, 3505 m, 22.8.1933 *Gill* 2054 (DD); Bashahr state, *Jorrie* 47229 (DD); Lahul Spiti, 30 km before Puh from Kaza, 3134 m, N 31°51.562′ E 78°36.808′, 8.8.2012, *Chaudhary* & Bajpai 259353 (LWG). **Jammu & Kashmir**: Ladakh, Chilung, 3353 m, July 1905, *Meebold* 1399 (CAL); Ladakh, Mulbekh-Shergul 3310–3200 m, 2.7.1976, *Wadhwa* 58792 (BSD); Ladakh, Dras, 3110 m, 17.7.1976, *Wadhwa* 58702 (BSD); Kashmir, Matayan, Dras, 3048 m, 1. 8.1891, *Gammie* (CAL, DD); Kashmir, Matayan, Dras, 3048 m, 1.8.1891, *Gammie* (DD); Dras,

3048 m, 22.7.1977, *Stainton* 7974 (BM); Dras, 3048 m, 7.6.1977, *Stainton* 7806 (BM). **Uttarakhand:** Uttarkashi, Bhairnghatti-Lanka, 3000 m, 27.5.1977, *Aswal & Party* 6478 (CDRI); Uttarkashi, Nelang Naga, 3800 m, July 1988, *Naithani* 86804 (BSD).

Notes: Ali (1958, 9161, 1977) has recorded it from India (Dras), however, Wenninger (1991) and Podlech & Zarre (2013) have not recorded it from India.

8. *Astragalus* sect. *Dissitiflori*

DC., Prodr. 2: 284. 1825; Podlech & Zarre, Tax. Rev. gen. *Astragalus* (Legum.) Old World 3: 1828. 2013. –Lectotype species: *Astragalus virgatus* Pall.

Herbs with well developed stems, hairy with medifixed or asymmetrically bifurcate or rarely subbasifixed warty hairs. Stipules free or adnate to petiole. Racemes peduncled, loose. Calyx tubular. Pods sessile or rarely stipitate, linear to narrowly oblong, glabrous or hairy, generally completely bilocular or half bilocular.

Key to the Species

1a. Peduncle c. 3 cm long; pods half bilocular .... **1. *A. anfractuosus***

1b. Peduncle 5–11 cm long; pods completly bilocular .... **2. *A. imitensis***

1. *Astragalus anfractuosus* Bunge, Mém. Acad. Imp. Sci. Saint Pétersbourg 11 (16): 125. 1868 & 15 (1): 128. 1869; Kumar & Sane, Legum. South Asia: Checkl. 222. 2003; Podlech & Zarre, Tax. Rev. gen. *Astragalus* (Legum.) Old World 3: 1853. 2013. –Type: Tibeto Occidentali, W of Leh, 12000 ft., *Hooker & Thomson* (sub *A. subulato*) (lectlo P *n. v.*; iso BM *n. v.*, C *n. v.*, CGE *n. v.*, K!, L *n. v.*, M *n. v.*, MSB *n. v.*, P *n. v.*, W *n. v.*).

A. subuliformis auct. non DC. 1802; Ali, Kew Bull. 1958: 316. 1958; Biologia 7: 77. 1961; Kew Bull. 28 (2): 303. 1973 & in Nasir & Ali, Fl. W. Pakistan 100: 211, f. 28 A–F. 1977; Sanjappa, Legum. India 95. 1992.

A. subulatus auct. non Pallas 1800 & non Desf. 1799; Baker in Hook. f., Fl. Brit. India 2: 136. 1876.

Plates 11, 51

Plants perennial herbs, prostrate, stems 2–3-branched (more or less dichotomously branched), slender, pubescent with appressed, medifixed, white hairs (young stem densely hairy). Stipules minute to 2.5 mm long, free or sometimes adnate to the petiole at base, linear-lanceolate, pubescent outside, hairs similar to stem but sometimes with mixed white and black hairs. Leaves 3–5 cm long, imparipinnately compound; petiole 9–13 mm long; rachis 13–20 mm long; leaflets 2–4 pairs, 7–8 × 1–2 mm, sparse, thick, linear, acute to subobtuse at apex, pubescent on both sides with appressed, white, medifixed hairs. Inflorescence axillary, peduncled raceme, 3–10- flowered; peduncle c. 25 mm long, exceeding the subtending leaf, hairy as stem. Bracts c. 1 mm long, persistent, lanceolate, hairy with appressed, mixed black and white hairs outside. Flowers c. 15 mm long, white; pedicel c. 1 mm long, hairy with appressed, mixed black and white hairs. Calyx 8–10 mm long, tubular, oblique at mouth, pubescent with appressed, mixed black and white hairs outside, tube 7–8 mm long, lobes 1–2 mm long, linear-lanceolate, hairy on both sides, hairs on inner side only white. Pods c. 25 × 2 mm, nodded, narrowly cylindrical, straight, minutely stalked, narrowed towards apex into a beak like structure, half bilocular, hairy with appressed, medifixed, white or mixed white and black hairs.

Flowering & Fruiting: July–August.

Habitat: Altitudes 3100–4100 m.

Distribution: INDIA: Himachal Pradesh, Jammu & Kashmir; CHINA; PAKISTAN.

Specimens examined: **Himachal Pradesh**: Lahul-Spiti, Takling, 4100 m, 27.7.1972, *Bhattacharyya* 49092 (BSD); Spiti valley, Near Quato, 3944 m, N 32°25.877′ E 77°55.054 2.8.2006, *Chaudhary et al.* 229520 (LWG); . Lahul-Spiti, 15 km after Lossar towards Kaza, 3934 m, N 32° 25.845′, E 77° 55.067′, 7.8.2012, *Chaudhary & Bajpai* 259346 (LWG). **Jammu & Kashmir**: Ladakh, Dras, 27.7.1988, *Chowdhery & Uniyal* 85823 (BSD); Ladakh, Nimu, 3100 m, 5. 8.1980, *Bhattacharyya* 71443 (BSD).

Notes: Baker (1876) identified this species as *A. subulatus* Pallas. Ali (1977) accepted *A. subuliformis* DC. as coreect name in broader concept and treated *A. subulatus* Pallas, *A. anfractuosus* Bunge and *A. suffulcatus* Bunge as synonyms under it. In the recent work (Kumar & Sane, 2003; Podlech & Zarre, 2013) *A. anfractuosus* Bunge, *A. suffulcatus* Bunge and *A. subuliformis* DC. (=*A. subulatus* Pallas) have been treate as distinct species. The Indian plants have been identified as *A. anfractuosus* Bunge.

2. *Astragalus imitensis* Ali, Candollea 17: 90, t. 14 & 17. 1960, non *A. imetensis* Boriss. 1947; Biologia 7: 79. 1961 & in Nasir & Ali, Fl. W. Pakistan 100: 211. 1977; Sanjappa, Legum. India 89. 1992. Kumar & Sane, Legum. South Asia: Checkl. 231. 2003; L.B.Chaudhary & S.K.Srivast., Taiwania 52 (1): 28. 2007; –Type: India, Kashmir, Imit, 2900 m, 2–3.8.1954, *Schmid* 2072 (holo G *n. v.*; iso RAW *n. v.*).

A. neo-schmidii Podlech & Rout in Fl. Iranica 178: 248. 2010; Podlech & Zarre, Tax. Rev. gen. *Astragalus* (Legum.) Old World 3: 1918. 2013., nom. illeg. nom. superfl.

Plants perennial herbs, erect, 30–40 cm high, caespitose, stems several, branched from the base, cylindrical, pilose with medifixed, appressed, white hairs, sometimes also with few black hairs. Stipules c. 2 mm. long, nearly free, subulate, white or occasionally mixed white and black hairy. Leaves imparipinnately compound, 3–5.5 cm long; petiole 1–2.5 cm long, hairy with medifixed, appressed, white or sometimes with few black hairs; leaflets 4–5 pairs, 5–10 × 2–4 mm, elliptical to obovate, rounded to subacute with minute mucro at apex, entire along margins, pilose on both sides with white hairs. Inflorescence 9–23 cm long, axillary, raceme, loosely 10–20-flowered; peduncle 5–18 cm long, densely hairy with medifixed, appressed, white or sometimes with few black hairs or sometimes glabrescent with age; floral rachis 4–5 cm long, strongly elongating up to 20 cm in fruits, hairy like peduncle but few black hairs mixed in. Bracts 1–2 mm long, membranous, narrowly triangular to ovate-triangular, sparsely white and black hairy. Flowers purplish-pink; pedicel c. 1 mm long, white and black hairy. Calyx c. 7 mm. long, tubular-campanulate, strongly oblique at mouth, densely covered with appressed white or white and black hairs, teeth 1.5–2 mm long, subulate, hairy on inner side. Standard 12–13 × 6 mm, rhomboid, retuse at apex; wing petals c. 11 × 1.5–2 mm, lamina narrowly oblong, emarginated at apex, claw 5–6 mm long; keel petals c. 8 × 2 mm, lamina triangular-obovate, incurved towards apex, claw c. 0.5 cm. long. Staminal tube truncate at apex. Ovary c. 4 mm long, sessile, hairy; style c. 5 mm. long, slightly curved at the tip, glabrous; stigma capitate, glabrous. Pods sessile, c. 17 × 2 mm, linear, straight, shortly acuminate, pilose with white or mixed white and black hairs, completely bilocular, c. 8-seeded.

Flowering & Fruiting: August.

Distribution: INDIA: Jammu & Kashmir (Imit-Baltistan); PAKISTAN.

Specimens examined: Nil.

Notes: Not examined in the present study as except type collections, no other collections are available from Jammu & Kashmir. The description of the species is based on Ali (1977) and Podlech & Zarre (2013). Podlech & Rout (2010) published *A. neo-schmidii* Podlech & Rout as new name for *A. imitensis* Ali, 1960 (from Imit, Kashmir), while the latter is a valid and legitimate name and not a later homonym of *A. imetensis* Boriss., 1947 (from Imet, Kyrgyzstan) (see IPNI).

9. *Astragalus* sect. *Galegiformes*

DC., Prodr. 2: 294. 1825; Podlech & Zarre, Tax. Rev. gen. *Astragalus* (Legum.) Old World 1: 212. 2013. –Type species: *Astragalus galegiformis* L.

A. sect. Diplotheca Bunge, Mém. Acad. Imp. Sci. Saint Pétersbourg 11 (16): 25. 1868. –Lectotype species: *Astragalus graveolens* Benth.

A. sect. Chlorostachys Bunge, Mém. Acad. Imp. Sci. Saint Pétersbourg 11 (16): 27. 1868; Wenninger, Mitt. Bot. Staatss. Munchen 30: 1.1991; L.B. Chaudhary & Z.H. Khan in Pandey *et al.*, Pl. Tax.: Adv. & Relev. 379. 2005. –Type species: *Astragalus chlorostachys* Lindl.

A. sect. Sesbanella Bunge, Mém. Acad. Imp. Sci. Saint Pétersbourg 11 (16): 4. 1868. –Lectotype species: *Astragalus sesbanoides* Benth.

Perennial herbs to shrubs, erect, caulescent, stems long, thick, hollow. Hairs basifixed (partly medifixed in vegetative parts of *A. trichocarpus*), white and black or dark brown. Stipules free from petiole, free from each other or occasionally connate opposite the petiole. Leaves imparipinnately compound. Racemes long pedunculate, loosely to densely many-flowered, generally elongating in fruit. Bracts persistent or caducous. Bracteoles present or absent. Flowers yellow or pink. Calyx campanulate to tubular, slightly obliquely gibbous at base, oblique at mouth, teeth unequal. Petals glabrous. Ovary shortly to long stipitate, glabrous or hary; style and stigma glabrous. Pods partially to completely bilocular, valves membranous.

Key to the Species

1a. Stipules almost fully connate throughout the length opposite the petiole and forming nearly an entire blade .... **10. *A. stipulatus***

1b. Stipules free from each other or shortly to distinctly connate 2

2a. All stipules on main stem distinctly connate opposite the petiole or before the petiole 3

2b. All stipules on main stem free from each other 5

3a. Bracts not protruding flower buds; calyx teeth minute; pod stipe almost equal to calyx .... **2. *A. concretus***

3b. Bracts protruding flower buds; calyx teeth up to 4 mm long; pod stipe shorter than or longer than calyx 4

4a. Leaflets glabrous above; flowers white or creamy yellow; pods stipe shorter than calyx .... **5. *A. isabellae***

4b. Leaflets glabrous to glabrescent above; flowers yellow; pods stipe longer than calyx .... **6. *A. khasianus***

5a. Plants with both basifixed and medifixed hairs .... **11. *A. trichocarpus***

5b. Plants either glabrous or only with basifixed hairs 6

6a. Stipules 19–50 mm, both distinctly unequal .... **4. *A. graveolens***

6b. Stipules 4–8 mm long, both equal 7

7a. Ovary and pods thinly to densely pubescent 8

7b. Ovary and pods glabrous 11

8a. Calyx 2.5 mm long, campanulate; keel petals shorter than standard .... **8. *A. nainitalensis***

8b. Calyx 5–6 mm long (up to 9 mm long in fruits), tubular; keel petals equal to standard 9

9a. Teeeth of calyx very minute (less than 1 mm long) .... **12. *A. uttaranchalensis***

9b. Teeeth of calyx 2–4 mm long 10

10a. Stipules hairy; leaflets pubescent on both sides; calyx lobes c. 2 mm long; ovary thinly hairy; young pods glabrescent with white, appressed hairs; pod stipe longer than calyx .... **3. *A. emodi***

10b. Stipules almost glabrous; leaflets almost glabrous above, pubescent below; calyx lobes c. 4 mm long; ovary hairy; pods pubescent with blackish-brown, half spreading hairs; pod stipe equal to calyx .... **7. *A. maxwellii***

11a. Leaflets sericeous on both sides; bracts 9–10 mm long, distinctly protruding flower buds; flowers pinkish-purple; calyx teeth up to 4 mm long; stigma papillose .... **9. *A. stewartii***

11b. Leaflets sparsely pubescent on both sides or sometimes glabrescent above; bracts c. 4 mm long, scarsely protruding flower buds; flowers yellowish-white; calyx teeth up to 1 mm long; stigma capitates .... **1. *A. chlorostachys***

1. *Astragalus chlorostachys* Lindl., Trans. Hort. Soc. 7: 249. 1830; Baker in Hook. f., Fl. Brit. India 2: 128. 1876; Ali, Biologia 7: 34. 1961 & in Nasir & Ali, Fl. W. Pakistan 100: 174. 1977; Grierson & Long in Grierson & Long, Fl. Bhutan 1(3): 716. 1987; Sanjappa, Legum. India 85. 1992; Wenninger, Mitt. Bot. Staatss. Munchen 30: 42. 1991; Podlech in Rech. f., Fl. Iranica 174: t. 41. 1999; Kumar & Sane, Legum. South Asia: Checkl. 225. 2003; L.B.Chaudhary & Z.H.Khan in Pandey *et al.*, Pl. Tax.: Adv. Relev. 381, f. 1. 2005; Podlech & Zarre, Tax. Rev. gen. *Astragalus* (Legum.) Old World 1: 221. 2013. –Type: Cultivated in Chiswich gardens, seeds received from Nepal, 1825 (holo CGE *n. v.*).

Tragacantha chlorostachys (Lindll.) O. Kuntze, Revis. Gen. 2: 944. 1891.

Hedysarum strobiliferum Baker in Hook. f., Fl. Brit. India 2: 146. 1876, non Linn. 1753. –Type: India, Kashmir, *Falconer s. n.* (holo *n. v.*).

Astragalus microdontus Baker, J. Linn. Soc. Bot. 18: 46. 1880. –Type: Pakistan, Kurram valley, Shendtoi, 7000–10000 ft., 1879, *Aitchison* 818 (lecto K!; iso G-BOIS *n. v.*).

Plates 12, 52

Plants woody herbs to under shrubs, 30–100 cm high, well branched, stems terete, smooth, hollow, sometimes partially reddish, pubescent or sometimes glabrous to glabrescent below and pubescent above or glabrous throughout, hairs appressed, forwardly oriented, white or occasionally intermixed with black in nodal region. Stipules 4–8 × 1.5–2.5 mm, greenish with purple tinge, free, spreading, lanceolate, acuminate, ciliate with spreading long hairs, glabrous to glabrescent outside, glabrous inside. Leaves 9–16 cm long, imparipinnately compound; petiole 1.5–2.7 cm long, sparsely pubescent; rachis 5–13.8 cm long, sparsely pubescent to densely tomentose; leaflets 5–9 pairs, 11–35 × 5–15 mm, opposite, subopposite or alternate, subsessile, oblong or oblong-ovate, subobtuse to cunneate at base, obtuse or rounded-apiculate at tip, sparsely pubescent on both sides, sometimes glabrescent above. Inflorescence axillary, long peduncled raceme, many-flowered, 7–30 cm long; peduncle 5–10 cm long, glabrescent to sparsely pubescent, hairs black in upper portion and white in lower portion or intermixed. Bracts c. 4 × 1.5 mm, scarcely exceeding the flower buds, caducous, tinged with purple colour, lanceolate, acuminate, sparsely ciliate along margins, glabrous or sometimes glabrescent outside. Flowers 11–13 mm long, yellowish-white or yellow, at maturity often become pinkish-violet, half nodded; pedicel 1.5–2 mm long, sparsely pubescent with black hairs. Calyx 4.5–6.5 mm long, persistent, tubular, tube 4–6 mm long, oblique at mouth, pubescent with appressed, black or sometimes intermixed with long, white hairs outside, glabrous inside, lobes minute to 1 mm long, unequal, longer on ventral side, pubescent on both sides. Standard 11–13 × 4.5–5 mm, narrowly obovate, emarginate, claw c. 2 mm long; wing petals 11.2–13 mm long, lamina c. 5.2 × 1.5 mm, narrowly ovate, subobtuse at apex, claw c. 6 mm long; keel petals c. 11 mm long, almost equal to standard and wing petals, lamina c. 4 × 2.5 mm, claw c. 7 mm long. Staminal sheath obtuse at apex. Ovary c. 3 mm long, linear, glabrous, with c. 5 mm long stipe; style c. 4 mm long; stigma capitate, glabrous. Pods 13–18 × c. 4 mm, oblong or narrowly ellipsoid, turgid, tapering at both ends, with 1.5–2 mm long beak, with faintly reticulate and membranous surface, glabrous, partially to completely bilocular, 6–10-seeded, stipe c. 7 mm long, longer than calyx. Seeds c. 3.7 × 2.5 mm, reniform, turgid, smooth, glabrous, brown with blackish spots.

Flowering & Fruiting: July–September.

Habitat: Grows along agricultural fields, in grassy lands or on mountain slopes in open as well as along forest sides. Altitudes (1500–) 1800–3600 (–4200) m.

Distribution: INDIA: Himachal Pradesh, Jammu & Kashmir, Uttarakhand; AFGHANISTAN; BHUTAN; CHINA; NEPAL; PAKISTAN.

Specimens examined: **Himachal Pradesh**: Kullu, Gulaba, 3000 m, 26.9.1983, *Aswal* 11224 (CDRI);Gulaba, 2600 m, 27.6.1984, *Chowdhery & party* 76044 (BSD); Gulaba-Marhi, 24.8.1981, *Aswal* 10638 (CDRI); Manali, before Rohtang, N 38° 24.427′ E 77°14.806′, 3814 m, 28.7.2006, *Chaudhary et al.* 229418, 229419 (LWG); Rohtang, 6 Km before Marhi from Manali side, 3.8.2003, *Chaudhary & Khan* 225001 (LWG); Lahul-Spiti: Khoksar, 16.9.2001, *Chaudhary* 220675 (LWG); Khoksar, 3150 m, 2.8.2002, *Chaudhary & Khan* 206738 (LWG); Khoksar, 3200 m, 18.7.1972, *Bhattacharyya* 48669 (BSD); Khoksar, 3200 m, 4.8.1970, *Bhattacharyya* 40555 (BSD); Khoksar, 3200 m, 2.8.1971, *Bhattacharyya* 45018 (BSD); Khoksar, 3200 m, 2.7.1973, *Bhattacharyya* 51713 (BSD); Khoksar, 3200 m, 11.7.1977, *Aswal* 6575 (CDRI); Lahul, *Stoliczka s. n.* (CAL); Sissu, 3200 m, 3.8.2003, *Chaudhary & Khan* 225009 (LWG); near Sissu, 13.9.2001, *Chaudhary* 220642 (LWG); Keylong, 3200 m, 2.8.1935, *Nath s. n.* (CAL); Keylong, Bhagabasin, 3100 m, 8.8.1971, *Bhattacharyya* 45197 (BSD); Keylong, 3353 m, 21.6.1941, *Bor* 14916 (DD); Keylong-Jispa, 3300 m, 27.8.1981, *Aswal* 10680 (CDRI); Chhatru, 8.8.1994, *Murti & Singh* 88171 (BSD); Dalhousie, 1830 m, 15.9.1874, Clarke 22499 (BM); Dalhousie, 2438 m, 17.9.1874, *Clarke* 22660C, 22660H (CAL); Dalhousie and Chamba, 2134 m, 18.8.1896, *Gammie* 18023 (DD); Chamba between Kilars & Sach, 2800 m, 20.7.1964, *Nair* 32694 (BSD); Baspa valley, Brua Khud, c. 2300 m, 1.6.1972, *Janardhanan* 47609 (BSD); Bashahr, Pangi village, 2900 m, 3.9.1963, *Nair* 30439 (BSD); Bashahr, Rakcham, 2700 m, 27.9.1964, *Nair* 34317 (BSD); Bashahr, Pangi village, 2900 m, 3.9.1963, *Nair* 30436 (BSD); Bashahr, Chitkul, 3300 m, 29.9.1964, *Nair* 34428 (BSD); Chitkul, 17.8.1973, *Janardhanan* 52765 (BSD); Sangla, 2 km ahead on way to Rakcham, 6.8.1973, *Janardhanan* 48500 (BSD); Chamba, Kalatop, 2338 m, 14.7.1963, *Malhotra & Nair* 27941 (BSD); Between Killar & Sach, 2800 m, 20.7.1964, *Nair* 32697 (BSD); Near Danchu, 2400 m, 26.6.74, *Wadhwa* 53160 (BSD); Kinnaur, Rakchum Dogri, 3400 m, 10.7.1999, *Jishtu et al.* 1918 (BSD); Simla, Mosobora, 15 miles from Simla, 25.9.1963, *Kaul & Party* 48250 (LWG); Simla, 1886, *Collett* 585 (CAL); Mahasu near Simla, 2591 m, 8.7.77, *Gamble* 4970 B (CAL); Simla, Kufri, 2600 m, 1.8.2002, *Chaudhary & Khan* 206733 (LWG). Rahla forest, 3000 m, 1. 8.1962, *Singh* 22997 (BSD).

Jammu & Kashmir: Jammu province, Sanasar, 8.7.54, *Abrol* 3887 (LWG); Banihal, 25.6.1955, *Kaul & party* 23356 (LWG); Banihal mountains, facing Banihal valley, 1850 m, 13.9.1958, *Rao* 7608 (BSD); Brari, ± 2134 m, 30.7.1956, *Abrol* 5094 (LWG); Tangmarg, along Firozgurnala, 29.6.1955, *Kaul & Party* 23494 (LWG); Verinag, 1.7.1963, *Kaul & Party* 92828 (LWG); Verinag, soil conservation area, 3.7.1963, *Kaul & Party* 92881 (LWG); Sauziea-Gulemaidan, 2500–3000 m, 19.9.85, *Naithani* 78397 (BSD); Tangmarg, 3630 m, 2.7.1956, *Rao* 363 (BSD); Bhadrawaha Podi, 19.8.1986, *Vohra & Naithani* 82855 (BSD); Batakut, 19.6.1960, *Malhotra* 12288 (BSD); Kistawar/ Nagin Siru, 1800-3850 m, 22.9.1958. *Rao* 7834 (BSD); Bhadrawaha, Seoj, 3000 m, Aug. 1986, Collector ? 83067 (BSD); Barwan Kishan ganga valley, 5.8.87, *Wadhwa et al.* 84923 (BSD); Sind valley, Gagangir, 2438 m, 27. 8. 1940, *Ludlow & Sherriff* 8038 (BM); Dunga gali, Aug. 1880, *Young s. n.* (BM); Pahlgam, 2286 m, 6. 9. 1876, *Clarke* 31127A (BM); above Pahlgam, 2591 m, 13. 8. 1956, *Polinin* 56/466 (BM); Liddar valley, above Pahlgam, 2438–2743 m, 4. 8. 1893, *Duthie* 13470 (BM). **Uttarakhand**: Almora, Bakhodiyar Rilkot, Johar, 3048–3353 m, 5.10.1950, *Awasthi* 1828 (LWG); Chamoli, Kedarnath on the way Chaura Bari, 19.9.1969, *Kapoor & Party* 70550 (LWG); Chamoli, Kedarnath, 3581 m, 29.8.1980, *Palni & Party* 11882; Chamoli, Sutal-Setal, 8.9.89, *Hajra* 87900 (BSD); Chamoli, Kedarnath, 6.7.78. *Rao* 58488 (BSD); Kedarnath, Mandakini valley, 3500 m, *Mehrotra* 4597 (CDRI); Chamoli, June 1979, *Naithani* 68159 (BSD); Kedarnath, Rambra, 3300 m, 26.9.1958, *Rau* 8652 (BSD); Rambara, 2800 m, 13.8.1968, *Rau* 38711 (BSD); Ghangaria, 3000 m, 21.8.1963, *Bhattacharyya* 29611 (BSD); Ghangaria, 3100 m, 16.8.1963, *Bhattacharya* 29435 (BSD); Valley of flowers, 3500 m, 19.8.1963, *Bhattacharyya* 29597 (BSD); Chakrata, Kanasor, 2000 m, 21.7.1968, *Mishra* 38175 (BSD); Chakrata Division, near Deoban, 2743 m, 7.9.1943, *Raizada* 18215 (DD); Chakrata, Near Deobon, 2743 m, 2.7.1951, *Awasthi* 2641 (LWG); Chakrata, Deoban, near guest house, 2800 m, 12. 8. 2001, *Chaudhary* 220615 (LWG); Mussoorie, October 1897, *Mackinnon s. n.* (CAL); Terhri-Garhwal, Rhudughera, 3048–3353 m, 18.7.1883, *Duthie* 989 (CAL, DD); Bhyundar valley, 3600 m, 4.10.1962, *Bhattacharya* 24344 (BSD); Uttarkashi, Kedarganga valley, 3600 m, 30.8.83, *Bhattacharyya* 74649 (BSD); On the way to Valley of flowers, 3200 m, 20.10.1963, *Rau* 31737, (BSD); Way to Jumma from Dunagiri, 3000 m, 23.8.1974. *Naithani* 54156 (BSD); Uttarkashi, Lanka-Bhaironghathi, 3000 m, 3.6.1983, *Aswal* 9279 (CDRI). Uttarkashi, Kedarganga valley, 3600 m, 30.8.83, *Bhattacharyya* 74649 (BSD); Uttarkashi, Kalkattidhar, 20.8.96, *Balodi* 92178 (BSD); Uttarkashi, on way to Har Ki Dun, *Balodi* 89536 (BSD); Tehri, Khankhaliadhar, 4000 m, 27.6.1981, *Goel* 72851 (BSD); Pithoragarh, Hunthor, Milam village, *Malhotra & Balodi* 77640 (BSD); Pithoragarh, Bakrillogar, 3500 m, 6.8.72, *Arora* 49675 (BSD); Tehri, Boniudiyar, 3500 m, 16.9.1979, *Goel* 66095 (BSD); Tehri-Garhwal, Gumgum Gadh, 2438–2743 m, 3.8.1883, *Duthie* 990 B (CAL); Tehri-Garhwal, Ganges valley, *Duthie* 1493 (CAL); Tehri-Garhwal, Nag Tiba, Aug., 1899, *Duthie* 23017 (CAL); Kumaon, Kutti, Yangti valley, 3353–3657 m, 30.7.1886, *Duthie* 5470 (CAL); Nainital, 2134 m, October 1905, *Meebold* 1398 (CAL); Nainital, 2438 m, 21.9.1919, *Osmaston* 1162 (DD).

Notes: The species exhibits variations chiefly in the height of plants, size of leaflets, density of indumentum, length of Inflorescence and number of flowers in each Inflorescence. Usually, the plants are glabrescent to pubescent with minute hairs, while in some plants collected from Kashmir region (*Kaul & party* 233313) and Deoban (*Awasthi* 2641), the hairs are very dense and tomentose type.

Earlier this species was reported from Assam (Wenninger, 1991) and Sikkim (Sanjappa, 1992), however, no specimens have been studied in the present study from these areas. The report from Assam by Wenninger (1991) is based on the specimen 'Assam (now in Arunachal Pradesh), Walong, 7250 ft., 10. 10.1950, *Kingdon-Ward* 20244- BM)' which I have identified as *A. concretus* Benth. (=*A. xiphocarpus* Benth. ex Bunge). Others like Baker (1876), Grierson & Long (1987), Wenninger (1991) and Podlech & Zarre (2013) have also not recorded it from Sikkim.

2. *Astragalus concretus* Benth. in Royle, Illustr. Bot. Himal. Mount. 199. 1835; Ali, Biologia 7: 36. 1961 p. p. & in Nasir & Ali, Fl. W. Pakistan 100: 175, f. 17 L. 1977 p. p. (excluding *A. isabellae* Dunn); Grierson & Long in Grierson & Long, Fl. Bhutan 1(3): 716. 1987; Wenninger, Mitt. Bot. Staatss. Munchen 30: 56, f. 3 I, j, 17 a–e. 1991; Sanjappa, Legum. India 86. 1992 p. p. (excluding *A. isabellae* Dunn); Press *et al.*, Annot. Checkl. Flow. Nepal. 166. 2000; Kumar & Sane, Legum. South Asia: Checkl. 226. 2003; L.B.Chaudhary & Z.H.Khan in Pandey *et al.*, Pl. Tax.: Adv. & Relev. 385, f. 1. 2005; Podlech & Zarre, Tax. Rev. gen. *Astragalus* (Legum.) Old World 1: 222. 2013. –Type: India, Emodo *ad* Gossainthan, *Wallich* Cat. 5931 (lecto K *n. v.*: hb Bentham: photo LWG!, iso BR *n. v.*, CGE *n. v.*, E *n. v.*, G *n. v.*, K *n. v.*: hb Bentham & hb Hooker: photo LWG!).

A. viciodes Grah. (in Wall. Cat. 5931. 1831–1832, *nom. nud.*) ex Baker in Hook. f., Fl. Brit. India 2: 131. 1876, non Ledeb. 1831.

A. xiphocarpus Benth. ex Bunge, Mém. Acad. Imp. Sci. Saint Pétersbourg 11 (16): 27.1868; Baker in Hook. f., Fl. Brit. India 2: 129. 1876; Sanjappa, Legum. India 96. 1992; L.B.Chaudhary & Z.H.Khan, Rheedea 13: 75, f. 2. 2003; L.B.Chaudhary & Z.H.Khan in Pandey *et al.*, Pl. Tax.: Adv. & Relev. 391, f. 1. 2005. –Type: India, Sikkim, Khasrou, 9000 ft., *Hooker s. n.* (lecto K !; iso: CGE *n. v.*, K !)

Plates 13, 53

Plants erect herbs or undershrubs, branched, stems up to 90 cm high, terete, hollow, glabrous below, thinly pubescent above with minute, appressed, flat, blackish-brown hairs. Stipules 6–20 × 2–5 mm, free from petiole, shortly connate, not membranous, oblong or lanceolate, thinly pilose with spreading, black hairs outside, glabrous inside. Leaves 8–17 cm long, imparipinnately compound; petiole 1–3 cm long; rachis 9–12.5 cm long, brownish in colour, thinly pilose with spreading, white or blackish-brown hairs; leaflets 11–15 pairs, 12–27 × 4–10.5 mm, subsessile, alternate, subopposite or opposite, oblong or oblong-ovate, obtuse or subtruncate-apiculate or retuse at apex, glabrous on upper surface, sparsely pilose with white, spreading hairs on lower surface. Inflorescence 6–10 cm long, shorter than subtending leaf, axillary, peduncled raceme, 10–20-flowered; peduncle 3–6.5 cm long, glabrous to thinly pubescent like stem; floral rachis 2–4.5 cm long, thinly pubescent with appressed, blackish-brown hairs. Bracts 3–6 × 1.5 mm, equal to or slightly longer than pedicel, deciduous, not membranous, lanceolate or narrowly lanceolate, not protruding flower buds, pilose outside with black hairs. Flowers 12–14 mm long, yellow; pedicel 3– 4 mm long, pubescent with blackish-brown hairs. Calyx 6–6.5 mm long, tubular, degenerate except basal portin at maturity, tube oblique at mouth, thinly pubescent with brown hairs outside, glabrous inside, lobes very minute. Standard c. 12 × 4 mm, equal to wing and keel petals, backwardly deflexed, elliptic-obovate, claw not distinct; wing petals c. 12 mm long, lamina c. 5 × 1 mm, oblong, obtuse at apex, claw c. 7 mm long; keel petals c. 12 mm long, lamina c. 5 × 3 mm, claw c. 7 mm long. Staminal sheath c. 9 mm long, free filaments 1–1.5 mm long ; vexillary filament c. 10 mm long, free from staminal sheath. Ovary c. 4 mm long, glabrous, c. 6-ovuled, with c. 5 mm long stipe; style c. 3 mm long, inwardly curved; stigma capitate, glabrous. Pods 14–45 × 3–5 mm, narrowly oblong, narrowly ovoid or ensiform, straight or slightly curved, gradually tapering at both ends, subturgid, subtorulose, one side green and another side brownish, stipitate, glabrous, completely bilocular, 8–12-seeded, stipe c. 6 mm long, generally folded, equal to calyx. Seeds c. 2.5 × 2 mm, reniform, subcompressed, one side depressed, brownish, smooth, glabrous.

Flowering & Fruiting: (June) July–August (–September–October).

Habitat: On hill slopes and along road sides among herbs and shrubs. Altitudes 2500–4000 m.

Distribution: INDIA: Arunachal Pradesh, Sikkim, W. Bengal-Darjeeling; BHUTAN; CHINA; NEPAL.

Specimens examined: **Arunachal Pradesh:** Walong, 2210 m, 10.10.1950, *Kingdon-Ward* 20244 (BM). **Sikkim**: Lachung valley, August 1892, *Gammie* (CAL); Kaljonimie, 3048 m, September 1888, *King's collector* (CAL); Tankra, 3353 m, 5.8.1892, *Gammie* 638 (CAL); Without locality, 8.10. 1868, *Kurz* (CAL); Chynojyrie, August 1887, *King's collector* (CAL); Bijan, 1888, *King's collector* (CAL); Gowathang, October 1908, *Ribu* (CAL); Without locality, 16.10.1868, *Kurz* (CAL); Kaljmie, 3048 m, September 1888, *King's collector* (CAL); Zakophyar, 11.7.1986, *Raju* 5893 (BSHC); 2 km before Lachen, 12.7.1986, *Raju* 5850 (BSHC); Lachen, near Zema, 3212 m, N 27° 48.029′ E88° 33.231′, 4. 9. 2015, *Chaudhary & Kushwaha* 259380 (LWG); Lachen, 2438 m, 7.7.1909, *Smith & Cave* 968 (CAL); Above Lachen, 2720 m, 16.7.1996, *Sinha & Long* 17631 (BSHC); Ahead Lachen towards Thangu, 3300 m, 12.8.2004, *Chaudhary* 225061 (LWG); Without locality 3048–3353 m, *Hooker s. n.* (CAL- Type section). **West Bengal**: Darjeeling: Sandakphu, 3048 m, October 1898, collector ? 11947 (CAL); Sandakphu, 3657 m, July 1905, *Cave s. n.* (LBG); Without locality, 3353 m, 12.10.1875, *Gamble* 119, 119c (CAL); Without locality, 2134 m, July 1882, *Gamble* 10347 (CAL); Sandakphu, 3353 m, July 1881, *Gamble* 9504 (CAL); Sandakphu, 3353 m, September 1898, coll.? 11777 (CAL); East of Phalut, 3048 m, 21.10.1904, *Burkill* 25320 (CAL); Singalila, Phalut, 3930 m, Sabargan, 3300 m, Sandakphu, 3970 m, 9.7.1969, *Hara et al. n.* (BM); Phalut, 3600 m, 4.8.1972, *Kanai et al.*727195 (BM); Tanglu, 3000 m, Gairibas, 2800 m, Kalapokhari, 3000 m, Sandakphu, 3600 m, 31.7.1972, *Kanai et al.*727199 (BM); Sandakphu, 3600 m, 1.8. 1972, *Kanai et al.*727196 (BM)

3. ***Astragalus emodi*** Steudel, Nomencl. Bot. ed. 2, 1: 161. 1840; Baker in Hook. f., Fl. Brit. India. 2: 130. 1876; Ali in Biologia 7: 44. 1961; Sanjappa, Legum. India 87. 1992; Press *et al.*, Annot. Checkl. Flow. Pl. Nepal 166. 2000; L.B.Chaudhary & Z.H.Khan, Phytotaxo. 4: 60, f. 3 & 4 G-R. 2004; L.B.Chaudhary & Z.H.Khan in Pandey *et al.*, Pl. Tax.: Adv. & Relev. 386, f. 1. 2005. –Type: India, Kumaon towards Emod, *Wallich* 5930 (lecto K-W *n. v.*; iso CGE *n. v.*, K!, LE *n. v.*).

A. podocarpus Grah. (in Wall. Cat. 5930. 1831–1832, *nom. nud.*) ex Benth. in Royle, Illustr. Bot. Himal. Mount. 199. 1835, non C. A. Mey. 1831.

Plants perennial herbs, erect, tall, stems finely striate, glabrescent in lower part, thinly pubescent in upper part with minute, appressed, black hairs. Stipules 6–8 mm long, free, spreading, linear, pilose outside with mixed white and black hairs, ciliate along margins. Leaves c. 11 cm long, imparipinnately compound; petiole c. 9 mm long; rachis c. 10 cm long, hairy as stem; leaflets 11–13 pairs, 10–16 × 4–5 mm, alternate to opposite, oblong or oblong-elliptic, acute at tip, pubescent on both sides with appressed, white hairs. Inflorescence c. 17 cm long, axillary, peduncled raceme; raceme many-flowered, close; peduncle c. 9 cm long, pubescent with black, appressed hairs, floral rachis c. 8 cm long, hairy as peduncle. Bracts c. 6 mm long, longer than pedicel, protruding beyond flower buds, caducous, linear-lanceolate, pilose outside, ciliate along margins with mixed white and brown hairs. Flowers c. 13 mm long, nodded at maturity, white (yellow in FBI); pedicel 1.5–2 mm long, pubescent with black hairs. Calyx 5–6 mm long, persistent, tubular, oblique at mouth, tube c. 4 mm long, pubescent with appressed, black or sometimes mixed with white hairs outside, glabrous inside, lobes linear, pilose with mixed black and white hairs on both sides, upper one minute to 1 mm long, lower one 2 (–3) mm long. Standard c. 13 mm long, lamina c. 8 × 4 mm, oblong, retuse at apex, claw c. 5 mm long; wing petals c. 13 mm long, lamina c. 6 × 2 mm, oblong, obtuse, slightly curved towards apex, upper auricle c. 2 × 1 mm, claw c. 7 mm long; keel petals c. 13 mm long, sometimes slightly shorter than standard and wing petals, lamina c. 5 × 3 mm, abruptly incurved, upper auricle c. 1 mm long, claw c. 8 mm long. Staminal sheath c. 11.5 mm long, free filaments 1.5–2 mm long; vexillary filament c. 12 mm long. Ovary c. 4 mm long, glabrous (thinly pubescent at maturity), with c. 5 mm long stipe; style c. 4 mm long, incurved above; stigma capitate, glabrous. Young Pods thinly pubescent with white, appressed hairs, stipe c. 6 mm long, longer than calyx. Mature pods not seen.

Flowering & Fruiting: July–August.

Habitat: Altitudes 2600–3600 m.

Distribution: INDIA: Jammu & Kashmir, Uttarakhand; NEPAL.

Specimens examined: **Jammu & Kashmir:** Ladakh, Zojila-Matyan, 3200 m, 3.8.1980, *Bhattacharyya* 71260 (BSD).); Barwan Kirhan Ganga Valley, 5.8.1987, *Wadhwa et al.* 84923 (BSD); Kunsalwan, 2362 m, 22.7.1876, *Clarke* 29428B (BM). **Uttarakhand:** Kumaon, towards Emod, *Wallich* 5930 (K).

Notes: In gross morphology it resembles with *A. chlorostachys* Lindl., but distinctly differs in bearing linear stipules and bracts, longer calyx lobes and thinly pubescent ovary and young pods. Baker (1876) describes pods as glabrous, however, Ali (1977) has correctly observed pubescent young pods after the study of type materials. It also differs from *A. maxwellii* Royle ex Benth. by protruded bracts, almost equal petals and longer pod stipe than calyx (Chaudhary & Khan 2005). However, it has been treated as a synonym under *A. maxwellii* Royle ex Benth. in Wenninger (1991), Kumar & Sane (2003) and Podlech & Zarre (2013).

4. ***Astragalus graveolens*** Buch.-Ham. (in Wall., Cat. 5929. 1831–32, *nom. nud.*) ex Benth. in Royle, Illustr. Bot. Himal. Mount. 199. 1935; Bunge, Mém. Acad. Imp. Sci. Saint Pétersbourg 11 (16): 26. 1868 & 12: 31. 1869; Baker in Hook.f., Fl. Brit. India 2: 131. 1876; Ali, Biologia 7: 36. 1961 & in Nasir & Ali, Fl. W. Pakistan 100: 177, f. 25 G. 1977; Wenninger, Mitt. Bot. Staatss. Munchen 30: 63, f. 5 d–g, 18 d–h, 26 d–e. 1991; Sanjappa, Legum. India 88. 1992; Podlech in Rech. f., Fl. Iranica 174: t. 45. 1999; Press *et al.*, Anno. Checkl. Flow. Pl. Nepal 166. 2000; Kumar & Sane, Legum. South Asia: Checkl. 230. 2003; Podlech & Zarre, Tax. Rev. gen. *Astragalus* (Legum.) Old World 1: 226. 2013. –Type: Nepal, Bank of Rapti, 5.5.1814, *Buchanan-Hamilton, Wallich* Cat. 5929A (holo K-W *n. v.*; iso E *n. v.*, G *n. v.*).

A. rotundifolius Benth. in Royle, Illustr. Bot. Himal. Mount. 199. 1835, non Willd. 1802. –Type: India, Oude *et* Deyra Doon, 1825, *Wallich* Cat. 5292B (lecto K !).

Plates 14, 54, 55

Plants perennial subshrubs, profusely branched, stems terete, smooth, glabrous. Stipules 19– 26 × 5–15 mm (up to 50 × 25 mm in J&K specimens), free, foliaceous, both unequal, distinctly reticulately-veined, ovate-lanceolate or broadly ovate particularly upper ones, usually cordate and with a large auricle at base, entire along margins, acuminate or subobtuse-mucronate (in upper ones) with few hairs at tip. Leaves 16–19 cm long, imparipinnately compound; petiole c. 2 cm long, glabrous; rachis 14–17 cm long, faintly grooved above, glabrous, only a few hairs present near the base of petiolules; leaflets 4–8 pairs, 18–28 × 13–17 mm, subsessile, opposite or subopposite, ovate, ovate-elliptic, broadly obovate or oblong-elliptic, cuneate at base, entire along margins, obtuse or rounded-mucronate with few hairs at tip, glabrous, midrib raised below, lateral veins prominent on both surfaces; petiolule c. 1 mm long, glabrous. Inflorescence 10–32 cm long, axillary, long peduncled raceme, many-flowered; peduncle c. 3.5–6 (–12) cm long, glabrous. Bracts 4.5–6 (–10–17) × 1–1.5 mm, persistent, membranous, linear or linear lanceolate, acuminate at tip with few black hairs, otherwise glabrous. Bracteoles 2, minute to c. 2 mm long, setaceous with few black hairs on tip and base, sometimes reduced and represented by a bunch of black hairs. Flowers 18–20 mm long, light yellow, nodded; pedicel c. 1 mm long, thickened and grooved in upper part, glabrous. Calyx 9–10 (–15) mm long, tubular, glabrous, tube 6–7 mm long, oblique at mouth, teeth unequal, 2 (–4) mm long dorsal ones, 3.5 (–7) mm long ventral ones, triangular lanceolate, with red tinged margins, faintly veined, at maturity the entire calyx becomes membranous and prominently veined. Standard (16–) 19 mm long, lamina (12–) 15 × c. 6 mm, deflexed, elliptic-obovate, obtuse-retuse at apex, claw c. 4 mm long; wing petals 18–19 mm long, subequal to standard, lamina 9–11 × c. 3.5 mm, ovate with subobtuse tip, upper auricle c. 1 mm long, spurs present, claw c. 9 mm long; keel petals 15–16 mm long, lamina c. 6.5 × 4 mm, curved on ventral margins, upper auricle and spurs present, claw 9.5–10 mm long. Staminal sheath c. 13.5 mm long, obtuse at tip, free filaments 2.5–4.5 mm long, alternately shorter & longer; vexillary filament c. 17 mm long, free from staminal sheath; anthers dorsifixed. Ovary 4–5 mm long, linear, flat, glabrous, 16–22-ovuled, with c. 10 mm long stipe; style c. 4

mm long, abruptly curved from base, glabrous; stigma capitate, glabrous. Pods 30–40 × 6–8 mm, oblong, subcompressed, straight, tapering at base, long stipitate, shortly acuminate with a short beak at tip, glabrous, membranous, finely veined, completely bilocular, 10–16-seeded, dehiscent, stipe 7–9 mm long, almost equaling the calyx. Seeds c. 3 × 2 mm, brown, reniform, subcompressed, glabrous, smooth.

Flowering & Fruiting: July–September (May–June in Jammu & Kashmir region).

Habitat: On open mountain slopes among herbs and shrubs. Altitudes 1600–3050 m.

Distribution: INDIA: Himachal Pradesh, Jammu & Kashmir, Uttarakhand; AFGHANISTAN; CHINA; NEPAL; PAKISTAN.

Specimens examined: **Himachal Pradesh**: Kulu, Manikaran, 3 km from Manikaran towards Phulga, 17.9.2001, *Chaudhary* 220701 (LWG); After 2 km from Manikaran, 1874 m, 19.7.2008, *Chaudhary et al.* 249612 (LWG); Jarsy, Parbatti valley, 1600 m, 23.5.1978, *Aswal & Party* 9205 (CDRI); Chamba, Kuarsi, 2438 m, 16.9.1896, *Gammie* 18592 (DD); Sarahan, 2400 m, 6.5.1962, *Nair* 21901 (BSD); Baspa valley, Kundralas, on way from Karchan to Brua, c. 2200 m, 1.6.1972, *Janardhanan* 47590 (BSD); Baspa valley, Kharougla, on the way from Sangla to Rakcham, 18.5.1972, *Janardhanan* 47307 (BSD); Kundralas, on way from Baspa valley, 2200 m, 1.6.1972, *Janardhanan* 47590 (CAL); Between Umi and Chini, 2400 m, 1. 6. 1962, *Nair* 22243 (BSD); Bashahr, Urni, 2300 m, 31.5.1962, *Nair* 22197 (BSD); Bashahr, Urni, 2200 m, 30.5.1962, *Nair* 22117 (BSD); Bashahr, between Pangi village and Jangi, 2575 m, 6.6.1962, *Nair* 22488 (BSD); Rarung, 2800 m, 4.9.1963, *Nair* 30495 (BSD); Simla, Bashahr state, December 1913, *Hole* 222 (CAL); Kinnaur, Poh, near old P.W.D. resthouse, c. 3050 m, 7.6.1972, *Janardhanan* 47809 (BSD); Kinnaur, Pangi 3 Km beyond P.W.D. resthouse c. 2825 m, 5.6.1972, *Janardhanan* 47715 (BSD); Kinnaur, Sangla, 2900 m, 29.5.1999, *Jishtu & Subramani* 2113 (BSD); Palli, 1980 m, 31.5.1890, *Lace* 211 (CAL); Pangi, Kunawar, 2743 m, 18.8.1847, *Thomson* 1938 (CAL). **Jammu & Kashmir**: Dachigame, Game Sanctuary, 22.5.1957, *Kaul & Party* 41675 (LWG); Dachigame, reserve forest, 27.6.1955, *Kaul & Party* 23264 (LWG); Baramula, 3.6.1955, *Kaul & Party* 23564 (LWG); Baramula, 4.6.1954, *Singh* 11398 (LWG); Kishanganga valley, Manahu, (Sarli Sachha) Jhing, 2134 m, 9.7.1907, *Singh* 1703 (DD); Uri, Cashmer, 1220 m, June, 1905, *Meebold* (CAL); Stchibal, 1829 m, May 1938, *Timins* 71 (BM); Without locality, 2743 m, 1915, *Prescott-Decie s. n.* (BM). **Uttarakhand**: Mussoorie, near Landour, 13.6.1942, *Raizada* (DD); Mussorie, 1524 m, September 1895, *Mackinnon* (CAL); Dehra Dun, October 1895, *Mackinnon* (CAL); Chamoli, Malan, *Hajra* 87566 (BSD); Garhwal, 1864, *Falconer* 419 (CAL); Tehri-Garhwal, 1220–1524 m, April 1881, *Duthie* 1564 (CAL).

5. ***Astragalus isabellae*** Dunn, Kew Bull. 1924: 384. 1924; Wenninger, Mitt. Bot. Staatss. Munchen 30: 73, f. 6 e, f, 21 a–c. 1991; Kumar & Sane, Legum. South Asia: Checkl. 231. 2003; L.B.Chaudhary & Z.H.Khan in Pandey *et al.*, Pl. Tax.: Adv. & Relev. 387, f. 1. 2005; L.B.Chaudhary *et al.*, Taiwania 52 (3): 218, f. 3. 2007; Podlech & Zarre, Tax. Rev. gen. *Astragalus* (Legum.) Old World 1: 229. 2013. –Type: N. W. India, Kashmir, Pahlgam, narrow Nullah to Dum Patthar, 6.9.1920, *Stewart & Stewart* 5972 (lecto K!).

A. concretus sensu Ali, Biologia 7: 36. 1961 p. p. & in Nasir & Ali, Fl. W. Pakistan 100: 175, f. 17 L. 1977 p. p.; Sanjappa, Legum. India 86. 1992 p. p.

Plants erect herbs, 30–80 cm high, stems terete, smooth, hollow, thinly to moderately pubescent with appressed, white, minute hairs in extreme lower portion, glabrous to glabrescent in upper portion with appressed, white or mixed white and black hairs especially at nodal region. Stipules c. 17 × 9–10 mm, yellowish-white, membranous, free from petiole, encircling the stem, connate up to the middle or more opposite the petiole and also infront of the petiole, ovate with acuminate tip, glabrous to thinly pubescent towards base and densely hairy towards tip, hairs black or brown, appressed. Leaves 9.5–18 cm long; petiole 1.6–4 cm long; rachis 6.5–12 cm long, glabrescent to glabrous; leaflets 7–9 pairs, 13–31 × 4–12 mm, narrowly-oblong, oblong or narrowly-ovate, cuneate at base, entire and thinly ciliate along margins, acute or subobtuse to obtuse with a minute mucro at tip, glabrescent to thinly pilose with appressed, white hairs below, glabrous above. Inflorescence 9–23 cm long, axillary, long peduncled raceme; peduncle 5.5–17 cm long, glabrous to glabrescent with black hair, floral rachis very thinly pubescent with blackish-brown hairs; raceme many-flowered, close. Bracts 10–14 × 2–4 mm, persistent, longer than pedicel, protruding beyond flower bud, lanceolate, membranous, with glabrous to glabrescent surface and ciliate margins and tip with spreading, black hairs. Bracteoles minute–2.5 mm long, present or absent, quite variable, filiform to linear, membranous, thinly hairy. Flowers c. 15 mm long, white or creamy yellow; pedicel c. 2 mm long, pubescent with black hairs. Calyx 9–10 mm long, tubular, tube c. 6 mm long, oblique at mouth, glabrescent to thinly hairy with appressed, black hairs outside, glabrous inside, lobes hairy on both sides with black hairs, linear, unequal, dorsal ones 1–2 mm long, ventral ones 2–4 mm long. Standard c. 15 mm long, almost equal to or slightly longer than wing and keel petals, lamina narrowly obovate, retuse at apex, c. 12 × 5 mm, claw c. 3 mm long; wing petals 14–15 mm long, lamina c. 7 x 2 mm, oblong with obtuse tip, slightly curved towards apex, claw c. 8 mm long; keel petals 14–15 mm long, lamina c. 6 × 3 mm, claw c. 9 mm long. Ovary 7–8 mm long, linear, glabrous, with c. 4 mm long stipe; style glabrous, incurved; stigma capitate, with some very minute projections. Pods 17–22 × 4.5–5.5 mm, narrowly ellipsoid, gooved dorsally, with 5–6 mm long stipe, long acuminate with a 6–8 mm long beak, bilocular, glabrous.

Flowering & Fruiting: July–Sepember.

Habitat: On poen rocky grassy hill slopes. Altitudes 3000–4000 m.

Distribution: INDIA (Jammu & Kashmir). Endemic.

Specimens examined: **Jammu & Kashmir**: near Sinthan Pass, 3657 m, 1.7.1939, *Ludlow* 166 (BM); Wardhwan, 3657 m, 27.7.1963, *Khan & Shah* 84449 (LWG); Lidder valley, above Kainmul, 3657 m, 21.7.1893, *Duthie* 13121 (CAL); Goorai, 2743 m, 20.7.1876, *Clarke* 29292 (K); Sonamarg, 2895 m, 27.7.1922, *Stewart* 7201 (K).

6. ***Astragalus khasianus*** Benth. ex Bunge, Mém. Acad. Imp. Sci. Saint Pétersbourg 11 (16): 27. 1868 & 15 (1): 32. 1869; Baker in Hook. f., Fl. Brit. India 2: 130. 1876; Wenninger, Mitt. Bot. Staatss. Munchen 30: 75, f. 6 g–h, 21 d–f. 1991; Sanjappa,

Legum. India 90. 1992; Kumar & Sane, Legum. South Asia: Chekl. 232. 2003; L.B.Chaudhary & Z.H.Khan, J. Jap. Bot. 80 (1): 52, f. 1–2. 2005; L.B.Chaudhary & Z.H.Khan in Pandey *et al.*, Pl. Tax.: Adv. & Relev. 388, f. 1. 2005; Podlech & Zarre, Tax. Rev. gen. *Astragalus* (Legum.) Old World 1: 230. 2013. –Type: India, Khasia, 6000–7000 ft., *Hooker* & *Thomson s. n* (lecto K!; iso BM *n. v.*, C *n. v.*, CAL!, CGE *n. v.*, E *n. v.*, G *n. v.*, GOET *n. v.*, K!, L *n. v.*, LD *n. v.*, M *n. v.*, SB *n. v.*, P *n. v.*, STU *n. v.*, W *n. v.*).

Plants perennial shrubs, erect, tall, much branched, stems terete, glabrescent in lower portion, sparsely pilose with appressed or half appressed, forwardly oriented, white hairs in upper portion. Stipules (8–9–) 18–21 × 4–6 mm, persistent, free from petiole, connate in lower portion before the petiole or completely free, spreading, lanceolate with acuminate tip, very faintly veined, glabrous to glabrescent outside, glabrous inside, ciliate along margins with white hairs. Leaves 6–9.5 cm long, imparipinnately compound; petiole 10–12 mm long; rachis c. 8 cm long; petiole and rachis hairy as stem; leaflets (9–) 13–14 pairs, 13–17 × c. 5 mm, oblong, obtuse-mucronate at tip, glabrous to glabrescent above, pilose with white hairs below; petiolule c. 1 mm long, pilose with white hairs. Inflorescence 11–20 cm long, axillary, peduncled raceme; raceme many-flowered, close, subsecunded; peduncle 7–8.5 cm long, hairy as stem, floral rachis (4–) 9–11 cm long, pilose with mixed white and brown, half appressed hairs. Bracts c. 8 × 1–1.5 mm, caducous, protruding beyond flower buds (more than double in length than flower buds), submembranous, linear-lanceolate, hairy only along margins and midrib region on dorsal side, hairs white. Flowers c. 16 mm long, yellow, nodded at maturity; pedicel c. 2.5 mm long, pilose with mixed brown and white hairs. Calyx 6.5–8 mm long, persistent, campanulate, tube c. 5 mm long, oblique at mouth, pubescent with mixed brown and white, half appressed hairs outside, glabrous inside, teeth hairy on both sides, shorter than tube, lower ones 1.5–3 mm long, upper ones c. 1 mm long. Standard c. 16 mm long, claw c. 5 mm long, lamina c. 11 × 8 mm, oblong, emarginate at tip; wing petals c. 16 mm long, claw c. 9 mm long, upper auricle c. 2 × 1 mm, lamina c. 7 × 2 mm, oblong with obtuse tip; keel petals c. 15 mm long, slightly shorter than standard and wing petals, claw c. 10 mm long, upper auricle c. 1.5 mm long, lamina c. 5 × 3 mm. Staminal sheath c. 14 mm long, obtuse at apex, free filaments 1–2 mm long, alternately shorter and longer; vexillary filament c. 15 mm long. Ovary c. 7 mm long, linear, glabrous; 12–14-ovuled, with c. 7 mm long stipe; style incurved, c. 3 mm long; stigma capitate, glabrous. Pods c. 20 × 5 mm, oblong, turgid, narrowed at both ends, with 2–4 mm long beak at apex, stipitate, grooved dorsally, glabrous, partially bilocular, c. 12-seeded, stipe 9–10 mm long, longer than calyx.

Flowering & Fruiting: July–September (–November).

Distribution: INDIA: Meghalaya; BHUTAN; CHINA; MYANMAR.

Specimens examined: **Meghalaya**: Shillong, 1768 m, 16.8.1885, *Clarke* 38853C (BM); Shillong, 1768 m, 16.8.1885, *Clarke* 38853 E (CAL-student Herb.); Khasia, Lailankote, 1830 m, 28.11.1871, *Clarke* 14783C (CAL); Khasia, Lailankote, 1830 m, 28.11.1871, *Clarke* 14783D (CAL); Khasia, *Hooker s. n.* (CAL-Type section).

7. ***Astragalus maxwellii*** Royle ex Benth. in Royle, Illustr. Bot. Himal. Mount. 199. 1835; Ali, Kew Bull. 1958: 308. 1958; Biologia 7: 45. 1961 & in Nasir & Ali, Fl. W. Pakistan 100: 181. 1977; Wenninger, Mitt. Bot. Staatssamml. Munchen 30: 79, f. 7 d-e, 22 c-e. 1991 p. p. (excluding *A. emodi* Steudel); Sanjappa, Legum. India 91. 1992; Kumar & Sane, Legum. South Asia: Checkl. 234. 2003 p. p. (excluding *A. emodi* Steudel); L.B.Chaudhary & Z.H.Khan in Pandey *et al.*, Pl. Tax.: Adv. & Relev. 389, f. 1. 2005; L.B.Chaudhary *et al.*, Taiwania 52 (3): 220, f. 5. 2007; Podlech & Zarre, Tax. Rev. gen. *Astragalus* (Legum.) Old World 1: 235. 2013 p. p. (excluding *A. emodi* Steudel). –Type: India, Kashmir, Chandow Pass, *Maxwell s. n.* (holo LIV *n. v.*: hb. Royle).

A. ciliolatus Benth. ex Bunge, Mém. Acad. Imp. Sci. Saint Pétersbourg 11 (16): 27. 1868; Baker in Hook. f., Fl. Brit. India 2: 129. 1876. –Type: **India,** Kashmir, Baltal, 9500 ft., 28.9.1848, *Thomson* (lecto K!; iso K!).

Plate 56

Plants perennial herbs, erect, 30–50 cm or more high, stems slender, terete, hollow, sparsely pubescent with appressed, forwardly oriented, minute, black hairs. Stipules 6–7 × 1–2 mm, persistent, free, spreading, rarely shortly connate opposite petiole, lanceolate or deltoid to narrowly lanceolate with acuminate tip, ciliate with black hairs along margins, almost glabrous on both sides. Leaves 9–10 cm long, imparipinnately compound, petiole 8–30 mm long; rachis 5–7 cm long; petiole and rachis thinly hairy like stem; leaflets 4–7 pairs, 19–25 × 7–10 mm, alternate, subopposite or opposite, shortly petiolulate, oblong or narrowly ovate, obtuse, subtruncate or retuse with a very minute mucro at apex, sparsely pubescent to glabrous above, hairy with white, spreading or half spreading hairs below, mixed black and white hairs on midrib; petiolules c.1 mm long, pubescent with half spreading black hairs. Inflorescence axillary, long peduncled raceme; raceme compact, many-flowered; peduncle (in fruits) 8–11 cm long, pubescent with minute, appressed, black and intermixed with few long white hairs; floral rachis (in fruits) 1.5–3.5 cm long, more hairy than peduncle. Bracts 4–8 mm long, membranous, deciduous, narrowly triangular, backish brown hairy. Bracteoles 2, 1–3 mm long, persistent, situated at the tip of pedicel, linear pubescent. Flowers yellow; pedicel 1–3 mm long, densely hairy, glabrescent in fruits. Calyx (in fruits) c. 9 mm long, persistent, tube c. 5 mm long, oblique at mouth, hairy with black and intermixed with some long white hairs outside, teeth hairy on both sides, lower ones c. 4 mm long, upper ones c. 2 mm long. Pods 20–25 × c. 6 mm, oblong-ovoid, straight or slightly curved, turgid, narrowed at both ends, with 4–7 mm long beak at apex, partially bilocular, c. 12-seeded, stipitate, densely pubescent with blackish brown, unequal, half spreading hairs, stipe c. 8 mm long, ± as long as calyx. Seeds c. 3.2 × 2.8 mm, blackish-brown, turgid, smooth, glabrous.

Flowering & Fruiting : June–September.

Habitat: Altitudes 2450–3500 m.

Distribution : INDIA: Jammu & Kashmir, Uttarakhand. Endemic.

Specimens examined: **Jammu & Kashmir:** Nichinai, 3505 m, 2.9.1956, *Polunin* 56/648 (BM); Zojila, *Young s. n.* (BM); Kishtwar, 2450 m, 14. 6. 1848, *Thomson s. n.* (K); Sonamarg, 3050 m, 5. 8. 1921, *Stewart s. n.* (K); Baltal, 27. 9. 1848, *Thomson s. n.* (K). **Uttarakhand**: Johar, Tola, 3400 m, 2. 7. 2004, GRS & GS 14313 (WII).

Notes: Podlech & Zarre (2013) have reported it from Nepal without citing any specimen, however, Chaudhary & Siwakoti (2015) have not recorded it from Nepal.

8. *Astragalus nainitalensis* L.B.Chaudhary, Taiwania 52 (3): 221, f. 6, 11c. 2007. –Type: India, Uttaranchal, Kumaon, Nainital, Date of collection, Collector & collection number not known, acc. no.113884 (holo CAL!: photo LWG!).

Plants erect, stems faintly striate, glabrous. Stipules 4–6 × 0.5–1 mm, free, spreading, narrowly lanceolate, ciliate along with margins with long, white, spreading hairs. Leaves 8–18 cm long; petiole 1.5–3.0 cm long; rachis and petioles thinly pilose with white, appressed hairs; leaflets 10–17 pairs, 9–24 × 5–9 mm, oblong or oblong-ovate, obtuse or cuneate at base, obtuse with a mucro at apex, glabrous above, sparsely pilose with white, appressed hairs below; petiolule c. 1 mm long. Inflorescence 9–12 cm long, axillary, peduncled raceme; peduncle 3.5–4.5 cm long, thinly pubescent with light brown, appressed hairs; raceme many-flowered, lax. Bracts 1.5–2.5 mm long, deciduous, narrowly lanceolate, hairy with mixed white and brown hairs. Flowers 7–8 mm long, nodded; pedicel c. 1 mm long, pubescent with appressed, brown hairs. Calyx c. 2.5 mm long, campanulate, persistent, pubescent outside, tube c. 2 mm long, teeth minute, unequal, shorter on dorsal side, linear-triangular. Standard c. 7 × 3 mm, obovate, retuse at apex, without distinc claw; wing petals c. 8 m long, longer than standard, lamina c. 5 × 1.5 mm, upper auricle hooked, claw c. 3 mm long; keel petals c. 6 mm long, shorter than standard and wing petals, lamina c. 3 × 2.5 mm, obovate, curved and obtuse at apex, upper auricle minute, claw c. 3 mm long. Staminal sheath c. 5 mm long, obtuse at apex, upper free filaments 1–1.5 mm long, alternately shorter and longer; vexillary filament c. 5 mm long;. Ovary c. 4 mm long, depressed brown hairy, with c. 1.5 mm long stipe; style c. 1.5 mm long, abruptly curved from base; stigma capitate, glabrous. Pods c. 9 × 6 mm, ovoid, stipitate, rounded or obtuse at base, narrow at apex with c. 3 mm long beak, thinly pubescent, faintly veined, stipe c. 3 mm long, longer than calyx.

*Distrib*ution: INDIA:Uttarakhand-Kumaon. Endemic.

Specimens examined: Known only from type specimen.

Notes: It is chiefly characterized by free stipules, lax raceme with nodded flowers, longer wing petals than standard, shorter keel petals than standard and wing petals and ovoid pods with longer stipe than calyx. Podlech & Zarre (2013). have missded this species to include in their work.

9. *Astragalus stewartii* Baker in Hook. f., Fl. Brit. India 2: 129. 1876; Ali, Biologia 7: 35. 1961 & in Nasir & Ali, Fl. W. Pakistan 100: 175, f. 17 S. 1977; Wenninger, Mitt. Bot. Staatss. Munchen 30: 84, f. 7 g–f, 22 f. 1991; Sanjappa, Legum. India 94. 1992; Kumar & Sane, Legum. South Asia: Checkl. 241. 2003; L.B.Chaudhary & Z.H.Khan in Pandey *et al.*, Pl. Tax.: Adv. & Relev. 390, f. 1. 2005; Podlech &

Zarre, Tax. Rev. gen. *Astragalus* (Legum.) Old World 1: 239. 2013. –Type: Pakistan, Punjab, Hazara, temperate region, 5000–9000 ft., *Stewart s. n.* (holo K!).

A. longicaulis Baker in Hook. f., Fl. Brit. India 2: 129. 1876, non Pomel 1874. –Type: India, Kashmir, *Falconer s. n.* (holo K!).

A. bakeri Ali, Kew Bull. 1958: 307. 1958; Biologia 7: 35. 1961 & in Nasir & Ali, Fl. Pakistan 100: 175, 17 R, 26 F–I. 1977; Sanjappa, Legum. India 85. 1992. –Type: that of *A. longicaulis* Baker.

Plate 57

Plants erect shrubs, pubescent or sericeous, hairs appressed, hairy with two types of hairs, shorter ones black or white, longer ones white. Stipules c. 7 mm long, free, spreading, linear-lanceolate/setaceous, ciliate along margins, sericeous with appressed, white or mixed white and black hairs outside, glabrous inside. Leaves 11–20.5 cm long; petiole 10–20 mm long; rachis 10–16 cm long; petiole nad rachis hairy as stem; leaflets 7–12 pairs, 16–30 × 5–14 mm, alternate (in lower portion) to opposite (in upper portion), narrowly oblong to oblong or narrowly ovate, entire along margins, obtuse, retuse or subtruncate with small mucro at apex, sericeous on both sides with appressed, white hairs. Inflorescence 12–15 cm long, axillary, long peduncled raceme; raceme many-flowered, compact; peduncle 5.5 – 11 cm long, hairy as stem. Bracts 9–10 × 1.5 mm, caducous, distinctly protruding flower buds, linear-lanceolate, ciliate along margins, sericeous with white, long, appressed or intermixed with small, appressed, black hairs. Bracteoles 2, 1–2 mm long, situated near the tip of pedicel, linear, white hairy. Flowers c. 12 mm long, pinkish-purple; pedicel c. 1.5 mm long, ascending, deflexed at maturity, pubescent with black or white, half appressed hairs. Calyx 7–8 mm long, campanulate-tubular, tube 4–5 mm long, pubescent with black, appressed hairs, intermixed with long white hairs particularly on ventral side, lobes unequal, linear, hairy on both sides (hairs white on outer surface and black in inner surface), dorsal ones c. 1 mm long, ventral ones 2–4 mm long. Standard c. 12 mm long, lamina c. 9 × 4 mm, obovate, minutely retuse, claw c. 3 mm long; wing petals c. 12.5 mm long, lamina c. 6 × 1.5 mm, oblong with obtuse tip, claw c. 7.5 mm long; keel petals c. 12 mm long, lamina c. 4 × 3 mm, claw c. 8 mm long. Ovary c. 3.5 mm long, 4–7-ovuled, glabrous, with c. 4 mm long stipe; style c. 5 mm long, glabrous, incurved in upper part; stigma papillose (very minute hairs like structure). Pods (after Podlech & Zarre, 2013) 12–20 × 6–8 mm, ovoid to ellipsoid with a 3–4 mm long beak and 6–8 mm long stipe, shallowly grooved dorsally, glabrous (or with few white hairs when young), partially to completely bilocular.

Flowering & Fruiting: July–September.

Distribution: INDIA: Jammu & Kashmir, Uttarakhand (?); PAKISTAN.

Specimens examined: **Jammu & Kashmir**: Jhelum valley, Karauli forest, Near Rampur, 1980 m, 5.7.1940, *Ludlow & Sherriff* 7726 (BM); Kunzalwan, 2286 m, 21.7.1876, *Clarke* 29381A (CAL- student herb.); Minimarg, 2743–3048 m, August 1905, *Meebold* 1364 (CAL).

Notes: In general appearance, *A. stewartii* Baker resembles with *A. chlorostachys* Lindl. and *A. emodi* Steud., but distinctly differs in the indumentum, stipule, calyx and pod. Ali (1961) has recorded this species from Kumaon (Uttarakhand) based on specimen 'Byans, near Kuti, *Duthie* 2811 (K)' which seems to be a different species. Podlech & Zarre (2013) have also not recorded this species from Uttarakhand.

10. *Astragalus stipulatus* D. Don, Bot. Mag. 50: t. 2380. 1823; Baker in Hook. f., Fl. Brit. India 2:131. 1876; Grierson & Long in Grierson & Long, Fl. Bhutan 1 (3): 715. 1987; Wenninger, Mitt. Bot. Staatss. Munchen 30:86, f. 8 a–c, 23 a–b, 26 f–g. 1991; Sanjappa, Legum. India 94. 1992; Press *et al.*, Anno. Checkl. Fl. Pl. Nepal 167. 2000; Kumar & Sane, Legum. South Asia: Checkl. 241. 2003; Podlech & Zarre, Tax. Rev. gen. *Astragalus* (Legum.) Old World 1: 240. 2013. –Type: Nepal, *Wallich* Cat. 5928 (lecto K-W; iso CGE *n. v.*, G *n. v.*, G-DC *n. v.*).

A. stipulates var. *phulchokiensis* H. Ohashi, Fl. E. Himalaya (3rd Rep.) 57. 1957; Kumar & Sane, Legum. South Asia: Checkl. 241. 2003; L.B. Chaudhary & Siwakoti, Int. J. Curr. Res. Biosci. Plant. Biol. 2 (4): 40. 2015. –Type: C. Nepal, Pulchoki, S. of Kathmandu, 1500-2200 m, 6.9.1972 (holo TI *n. v.*; iso BM *n. v.*, KYO *n. v.*, TI *n. v.*).

Plates 15, 58

Plants perennial herbs to shrubs, erect, 0.5–2 m high, stems stout, hollow, sometimes partially reddish, glabrous. Stipules 1.7–5 × 0.9–2.5 cm, green, foliaceous, large, leaf opposed, free from petiole, amplexicaul, completely connate except tip portion on the opposite side of petiole, sometimes shortly connate infront of petiole, glabrous, ciliate with white hairs along margins. Leaves 20–40 cm long, imparipinnately compound; petiole 4–5 cm long, glabrous; rachis pilose with long white hairs in young condition, glabrous at maturity; leaflets 15–25 pairs, 13–43 × 3.5–11 mm, generally opposite, oblong or occasionally narrowly ovate or oblanceolate, subobtuse, obtuse or retuse with a fine mucro at apex, glabrous to thinly pilose above, pilose below with white, appressed hairs; petiolule c. 1.5 mm long, pilose with white hairs. Inflorescence axillary, peduncled raceme, 15–22 cm long (elongating in fruits), with numerous close flowers; peduncle 5–6 cm long, glabrous. Bracts c. 15 × 1 mm, membranous, narrowly lanceolate, acuminate at tip, ciliate with long, mixed black and white hairs. Flowers 13–15 mm long, light yellow; pedicel 3–4 mm long, densely pilose with blackish-brown, half spreading hairs. Calyx 11–13 mm long, tubular, persistent, densely pilose with black or blackish-brown, spreading hairs, tube 6–7 mm long, almost equal to teeth, teeth 5–6 mm long, linear-lanceolate, hairy on both sides. Standard c. 13 mm long, lamina c. 8 × 6 mm, ovate, claw c. 5 mm long; wing petals c. 13 mm long, lamina c. 6 × 1 mm, linear-oblong, obtuse at apex, upper auricle c. 2 × 1 mm, oblong with subacute tip, claw c. 7 mm long; keel petals c. 12.5 mm long, lamina c. 4.5 × 2 mm, oblong, abruptly curved above, upper auricle minute, calw c. 8 mm long. Staminal sheath c. 11.5 mm long, obtuse at apex, free filaments 1–2 mm long, alternately short and long; vexillary filament c. 12 mm long. Ovary c. 11 mm long, glabrous, with c. 1.5 mm long stipe; style c. 2 mm long, curved from base, filiform; stigma capitate, glabrous. Pods 25–35 × 5–7 mm, narrowly ovoid to narrowly oblong, grooved dorsally, tapering to the apex into a long beak, shortly stipitate (stipe 1–2 mm long), glabrous, bilocular, 12–16-seeded. Seeds c. 2 × 1.5 mm, reniform, subcompressed, brown, glabrous, smooth.

Flowering & Fruiting: August–October (–November).

Habitat: Occasionally or thinly scattered on hill slopes along road sides among grasses and shrubs under forest shade. Altitudes (1200–) 2000–3000 m.

Distribution: INDIA: Sikkim, West Bengal-Drjeeling; BHUTAN; CHINA; NEPAL.

Specimens examined: **Sikkim**: North Sikkim, On way to Lachen, 18.8.1989, *Srivastava* 10213 (BSHC); Lachen, 2438 m, September–October, 1909, *Ribu & Rhomoo* 3022 (CAL); Lachen valley, August 1892, *Gammie* (CAL); Lachen, 2134 m, *Smith & Cave* 2609 (CAL); Lachen, Between Lachen & Thangu, 2906 m, N27° 45.603′ E88° 32.388′, 4. 9. 2015, *Chaudhary & Kushwaha* 259375 (LWG); Lachung, 2743 m, 27.8.1892, *Gammie* 1013 (CAL); Lachung, 2743 m, 27.8.1892, *Gammie* 1013 (CAL); Between Chungthang and Lachung, after c. 11 km from Chungthang, 2400 m, 16.8.2004, *Chaudhary* 225085, 225088 (LWG); Between Chungthang and Lachung, 15.8.2004, *Chaudhary* 225076 (LWG); Lachung, before township, 2900 m, 15.8.2004, *Chaudhary* 225077 (LWG); Chungthang to Lachen, 26.7.1986, *Raju & Singh* 6309 (BSHC); Bansoi to Chaten, 18.9.1994, *Pradhan* 17158 (BSHC); Bok, 2134 m, 7.11.1910, *Ribu & Rhomoo* 4763 (CAL); without locality, *S. K. Mukherji* 2031 (CAL); without locality, 1524 m, *Kurz* (CAL). **West Bengal**: Darjeeling, Lingdam, 1220 m, 12.10.1870, *Clarke* 13050 (BM); Darjeeling, Dantan, 1830 m, 03.1.1911, *Ribu & Rhomoo* 4910 (CAL).

11. *Astragalus trichocarpus* Grah. (in Wall., Cat. 5296. 1831–32, *nom. nud.*) ex Benth. in Royle, Illustr. Bot. Himal. Mount. 199. 1835; Baker in Hook. f., Fl. Brit. India 2: 121. 1876; Ali, Biologia 7: 71. 1961 & in Nasir & Ali, Fl. W. Pakistan 100: 202, f. 27 A–E. 1977; Wenninger, Mitt. Bot. Staatss. Munchen 30: 88, f. 8 d–e, 23 c, 27 d–f. 1991; Sanjappa, Legum. India 96. 1992; Kumar & Sane, Legum. South Asia: Checkl. 243. 2003; L.B.Chaudhary *et al.*, Taiwania 52 (3): 224, f. 10 & 11G. 2007; Podlech & Zarre, Tax. Rev. gen. *Astragalus* (Legum.) Old World 1: 241. 2013. –Type: India, Himalaya, Himachal Pradesh, Sirmore, *Govan s. n.* (lecto K-*Wallich* 5926 *n. v.*; iso K!).

A. sesbanoides Royle ex Benth. in Royle, Illustr. Bot. Himal. Mount. 199. 1835. –Type: India, Mussooree, *Royle s. n.* (lecto LIV *n. v.*; iso K!, LIV *n. v.*).

Plants perennial herbs, erect, c. 65 cm or more high, branched, stems furrowed, pubescent with minute, white, appressed hairs. Stipules minute, narrowly triangular to subulate, free, glabrous. Leaves 5–10 cm long, imparipinnately compound, subsessile; leaflets 12–15 pairs, 3–9 × c. 2 mm, subsessile, opposite to alternate, oblong, subobtuse to obtuse, with a very minute mucro at apex, glabrous above, thinly pubescent with minute, white, appressed hairs below, midrib densely pubescent. Inflorescence 10–13 cm long, almost equal to or longer than subtending leaf, axillary, peduncled raceme; raceme many-flowered, laxed; peduncle 1.5–3 cm long, glabrescent. Bracts c. 1.5 mm long, linear-lanceolate, shorter than pedicel, pubescent outside. Flowers c. 10 mm long, pinkish-white or lilac; pedicel c. 2 mm long, pubescent with brown, minute, appressed hairs. Calyx c. 2.5 mm long, campanulate, tube oblique at mouth, glabrescent to glabrous, lobes minute, deltoid, pubescent. Standard c. 10 × 6 mm, equal to keel petals, rounded, emarginated at apex, prominently veined; wing petals c. 7 mm long, shorter than keel petals, lamina c. 5 × 1 mm, lanceolate, claw 2–3 mm long; keel petals c. 10 mm long. Ovary c. 5 mm long, densely white pubescent, with c. 3 mm long hairy stipe; style incurved above; stigma penicillate. Pods 7–11 × 3–4 mm, oblong, turgid, grooved dorsally, shortly acuminate at tip, densely pilose with white, half spreading hairs, bilocular, 4–6-seeded, stipe 4–5 mm long, longer than calyx, hairy.

Flowering & Fruiting: April–June (–August).

Habitat: On open grassy hill slopes. Altitudes 1500–2400 m.

Distribution: INDIA: Himachal Pradesh, Uttarakhand. Endemic.

Specimens examined: **Himachal Pradesh**: Simla, Chail, 15.8.1948, *Ram s. n.* (DD); Simla, 1829 m, 17.5.1877, *Gamble* 4243 B (CAL); Simla, Armandole, 1829 m, 22.8.1878, *Gamble* 5891C (CAL); Simla; Mahasu, 2134 m, 7.6.1906, *Burkill* 26824 (CAL); Simla, 17.5.1877, *Gamble* 4243 C (CAL); Manali, 1829 m, *Bor s. n.* (DD); Bashahr State, Bali, 2400 m, 11.5.1928, *Parker* 2823 (DD); Bashahr, Kotgarh to Nirt, 24.4.1891, *Lace* 811 (CAL). **Uttarakhand**: Jaunsar Bahar, near Kinani Pani, *Duthie* 1080 (DD); Jaunsar, Chakrata, 2134 m, April 1891, *Gamble* 22857 (CAL); Deoban range, 2438–2743 m, May 1887, collector ? *s. n.* (CAL); Tons valley, 1524–1829 m, 4.6.1894, *Duthie* 14586 (CAL, DD); Tons valley, above Maindrath, 1524 m, 2.5.1899, *Duthie* 22538 (CAL); near Mussorie, 1869, *King* (CAL); Mussoorie, June 1894, *Robson* 1 (CAL); Mussoorie range, 1869, collector? *s. n.* (CAL); Mussoorie, 4.4.1896, *Macknon s. n.* (CAL); Mussoorie, collector? *s. n.* (CAL); Mussoorie, 1829–2134 m, June 1877, *Duthie* 142 (CAL); Mussoorie-Chamba, Kaddu Khal, June 2003, *Singh* 90972 (BSD); Mussoorie, on way to Hathi Paon, 24. 4. 1961, *Saxena* 1866 (DD); Kolegurh, *Stoliczka s. n.* (CAL); Ranikhet, 1800 m, April 1939, *Legge* 57 (K).

Notes: Now very occasional in Mussoorie area. During recent period only one collection *Singh* 90972 (BSD) has been made. I could see only a few plants in Hathi Paon (Massoorie) area.

12. *Astragalus uttaranchalensis* L.B. Chaudhary & Z.H. Khan, Taiwania 51 (1): 36, f. 1 & 2. 2006; L.B. Chaudhary *et al.*, Taiwania 52 (3): 224. 2007. –Type: India, Uttaranchal, Uttarkashi, Chirbasa, 3500 m, 5.9.2002, *Khan* 223088 (holo LWG!; iso LWG!, CAL!, BSD!).

Plates 16, 59

Plants perennial herbs, erect, 60–90 cm high, gregarious, stems terete, hollow, branched, pubescent with minute, appressed or spreading, white or mixed black and white hairs. Stipules 4–8 × 2–3 mm, free, spreading or deflexed, linear-triangular to triangular with acuminate tip, pilose with spreading, white or mixed black and white hairs outside, glabrous inside, ciliate along margins. Leaves 10–15 cm long, imparipinnately compoundly compound; petiole 1.5–3 cm long; rachis 8.5–11 cm long, pilose with white, spreading hairs or pubescent with appressed, mixed black and white hairs; leaflets 9–12 pairs, 18–28 × 4–11 mm, alternate, subopposite or opposite, oblong or narrowly ovate, entire along margins, obtuse, truncate or retuse with a fine mucro at apex, glabrescent to thinly pilose above, pilose with spreading, white hairs below. Inflorescence axillary, peduncled raceme, 4.5–15 cm long, shorter than subtending leaf; peduncle 2.5–9 cm long, pubescent with spreading, white hairs or appressed, black hairs; raceme 2–6 cm long, 10–20-flowered, close. Bracts 4–5 × c. 1 mm, not protruded beyond flower buds, almost equal to or slightly longer than pedicel, deciduous, ovate, lanceolate or narrowly lanceolate, pilose with mixed black and white hairs outside, glabrous inside, ciliate along margins. Flowers 11–12 mm long, yellowish-white, becoming purple at maturity; pedicel 2–3.5 mm long, pubescent with black or brown, spreading hairs. Calyx 4–5 mm long, tubular, persistent, oblique at mouth, pilose with black or brown, appressed hairs outside, glabrous inside, teeth very minute (less than 1 mm long). Standard c. 11 mm long, lamina c. 8 × 6 mm, oblong-elliptic with retuse apex, claw c. 3 mm long; wing petals 11–12 mm long, lamina c. 6 × 2 mm, oblong, obtuse at apex, upper auricle c. 2 × 1 mm, claw c. 6 mm long; keel petals 10–11 mm long, lamina c. 4 × 3 mm, upper auricle c. 1 mm long, claw c. 6 mm long. Staminal sheath c.10 mm long, obtuse at apex, free filaments 1–2 mm long, alternately shorter and longer; vexillary filament c. 11 mm long, free. Ovary c. 4 mm long, sparsely to densely pubescent with appressed, white hairs, with c. 5 mm long stipe; style gradually curved from the base, glabrous; stigma capitate, very finely penicillate. Young pods pubescent with appressed, white hairs. Pods 14–22 × 3–5 mm, partially reddish, oblong, straight or slightly falcate, turgid, narrowed at both ends, with 2–4 mm long apical beak and 6–7 mm long stipe stipitate, thinly pubescent to glabrescent with appressed, white hairs (or occasionally becoming glabrous at full maturity), partially to completely bilocular, 2–5-seeded. Seeds 2.5–3 × 1.5–2 mm, reniform, subturgid to turgid, brown or brownish-black, smooth, glabrous.

Flowering & Fruiting : July–September.

Habitat: Hill slopes among herbs and shrubs at altitudes ranging from 3200–3500 m elevations.

Distribution: INDIA: Uttarakhand. Endemic.

Specimens examined: **Uttarakhand:** Garhwal, Uttarkashi: Above Gangotri, 3137 m, N 30° 59.725'E 78°56.975', 4.9.2006, *Chaudhary et al.* 229550 (LWG); Gangotri, after 2 km towards Chirbasa, 3200 m, 5.9.2002, *Khan* 223078 (LWG); Gangotri to Chirbasa , 3277 m, N 30° 59.288'E 79° 00.081', 4.9.2006, *Chaudhary et al.* 229553 (LWG); Chirbasa, 3500 m, 5.9.2002, *Khan* 223089, 223094, 223097 (LWG); Chirbasa to Bhojwasa, 3314 m, N30° 58.827' E 79° 01.855', 4.9.2006, *Chaudhary et al.* 229549 (LWG); Approaching Bhojwasa, 3700 m, 4.9.1983, *Bhattacharyya* 74873 (BSD); Bhojwasa-Gangotri, 31.7.1987, *Uniyal & Balodi* 880538 (BSD); Kumaon, Gori valley: Tola, 15.8.1900, *Inayat* 24307 (CAL); Parbhu, 12.8.1900, *Inayat* 24309 (CAL).

Notes: Podlech & Zarre (2013.) have missded this species to include in their work.

10. *Astragalus* sect. *Helmia*

Bunge, Mém. Acad. Imp. Sci. Saint Pétersbourg 11 (16): 111. 1868; Podlech & Zarre, Tax. Rev. gen. *Astragalus* (Legum.) Old World 3: 2025. 2013. –Type species: *Astragalus helmii* Fisch. ex DC.

A. sect. *Chomutoviana* B. Fedtsch., Beih. Bot. Centralbl 23 (2): 367. 1908. –Type species: *Astragalus chomutovii* B. Fedtsch.

Perennial herbs, dwarf, stemless or with very short stems, hairy with bifurcate hairs. Stipules connate opposite the petiole. Racemes peduncled, few to about 20-flowered. Calyx campanulate to tubular. Petals yellow or purple. Pods sessile to subsessile, inflated or compressed laterally, keeled ventrally, narrowly grooved to flattened dorsally, unilocular to bilocular, densely covered with spreading hairs, rarely glabrous.

1. ***Astragalus arnoldii*** Hemsl. & H. Pearson, J. Linn. Soc., Bot. 35: 172. 1902; Ohba *et al.*, J. Jap. Bot. 70: 14, f. 1B. 1994; L.B.Chaudhary & S.K.Srivast., Taiwania 52: 26, f. 2. 2007; Podlech & Zarre, Tax. Rev. gen. *Astragalus* (Legum.) Old World 3: 2028. 2013. –Type: Tibet, Game, 17500 ft., 24.6.1896, *Deasy* & *Pike* 810 (lecto K!).

A. orotrephes W. W. Smith, Trans. Bot. Soc. Edinburgh 26: 178. 1913; L.B.Chaudhary, Rheedea 15: 78, f. 1. 2005. –Type: India, East Himalaya, Sikkim, July 1902, *Political Office, Sikkim* (holo CAL!: photo LWG; iso K *n. v.*).

Plants perennial herbs, dwarf, only c. 5 cm above the ground, very densely caespitose; rootstock rhizomatous, c. 15 cm long, thick and woody, stems numerous, very short, densely covered with leaves, very compact in hemispherical form. Stipules 4–5 × c. 1.2 mm, membranous, ovate with acuminate tip, adnate to the petiole up to the middle, shortly connate opposite the petiole, pilose to densely pilose with white, long, appressed hairs outside, glabrous inside. Leaves c. 9 mm long, imparipinnately compound; petiole c. 4 mm long; rachis c. 4 mm long, densely sericeous with white, half spreading hairs; leaflets c. 7 pairs, 1.5–3 mm long, close, sessile, ovate or obovate-elliptic , densely sericeous with white, long hairs on both sides. Flowers solitary on very short peduncle, c. 5 mm long. Bracts c. 2 mm long, ovate with acuminate tip, pilose with white, long hairs outside, glabrous inside. Calyx c. 4 mm long, tubular-campanulate, tube c. 2.5 mm long, densely hairy outside with white, long, appressed hairs, teeth slightly shorter than tube, linear lanceolate, densely hairy with mixed black and white hairs both sides. Standard c. 6 × 3 mm, nearly orbicular with emarginated apex; wing petals c. 5 mm long; keel petals c. 4.5 mm long.; Staminal sheath c. 3 mm long, truncate at apex, free filaments c. 2 mm long; vexillary filament c. 4 mm long. Ovary c. 2 mm long, sessile, densely white hairy; style c. 2.5 mm long. Pods 5–7 mm long, sessile, ovoid with a 1 mm long beak, inflated, densely hairy with white, long hairs, unilocular.

Flowering & Fruiting: June.

Distribution: INDIA: Jammu & Kashmir, Sikkim, CHINA; PAKISTAN.

Specimens examined: **Jammu & Kashmir**: Ladakh, Top of Taklang pass, 26. 6. 1856, collector ? 1645 (BM: photo LWG). **East Himalaya:** July 1902, *Political Office, Sikkim* (CAL).

Notes: A very rare species in India. Podlech & Zarre (2013) have recorded it from Nepal without citing any specimen.

11. *Astragalus* sect. *Herpocaulos*

Bunge, Mém. Acad. Imp. Sci. Saint Pétersbourg 11 (16): 9. 1868; Podlech & Zarre, Tax. Rev. gen. *Astragalus* (Legum.) Old World 1: 167. 2013. –Lectotype species: *Astragalus arabicus* Bunge.

Annual herbs, caulescent, covered with bifurcate hairs. Stipules minute, free. Racemes peduncled, capitate or cylindric. Bracts minute, persisitent. Calyx campanulate. Petals slightly exceeding the calyx. Pods sessile, narrowly ellipsoid, grooved dorsally, unilocular.

1. ***Astragalus vogelii*** (Webb) Bornm. subsp. ***fatimensis*** Maire, Mém. Soc. Hist. Nat. Afr. Nord 3: 126. 1933; Kumar & Sane, Legum. South Asia: Checkl. 244. 2003. –Type: Arabia, El Gidon, 1825, *Ehrenberg s. n.* (lecto P *n. v.*; iso BR *n. v.*, C *n. v.*, K!, L *n. v.*, LE *n. v.*, MSB *n. v.*, P *n. v.*).

Podlechiella vogelii (Webb) Maassoumi & Kaz. Osaloo subsp. *fatimensis* (Maire) Maassoumi & Kaz. Osaloo, Plant Syst. Evol. 242: 22. 2003; L.B.Chaudhary & S.K.Srivast., Taiwania 52 (1): 32. 2007; Podlech & Zarre, Tax. Rev. gen. *Astragalus* (Legum.) Old World 1: 169. 2013.

Astragalus arabicus auct. non Kotschy 1866; Bunge, Mém. Acad. Imp. Sci. Saint Pétersbourg 11 (16): 9. 1868.

A. prolixus auct. non Sieb. ex Bunge 1868; Baker in Hook. f., Fl. Brit. India 2: 121. 1876; Cooke, Fl. Pres. Bombay 1: 329. 1903.

A. fatimensis Hochst. ex Blatter, Rec. Bot. Surv. India 8 (1): 156. 1921; Ali, Biologia 7: 79. 1961 & in Nasir & Ali, Fl. W. Pakistan 100: 209. 1977; Sanjappa, Legum. India 87. 1992. –Type: India, Punjab, *Edgeworth s. n.* (holo K *n. v.*).

Plants annual herbs, prostrate, caespitose, stems slender, c. 7 cm long, pilose with appressed, medifixed, white hairs. Stipules 1.5–2 mm, free or slightly adnate with petiole at base, triangular-lanceolate, pilose, with white, medifixed hairs outside, glabrous inside, ciliate along margins. Leaves 18–22 mm long, imparipinnately compound; petiole c. 6 mm long; rachis 9–12 mm long, hairy as stem; leaflets 3–7 pairs, 3–5 × 1–2 mm, subsessile, ovate or oblong, obtuse or minutely retuse at apex, pilose on both sides with appressed, medifixed, white hairs. Inflorescence axillary, shortly peduncled umbel, 2–3-flowered; peduncle 2–5 mm long, hairy as stem. Bracts c.1 mm long, longer than pedicel, pilose with white hairs. Flowers c. 3 mm long, subsessile; pedicel less than 1 mm long. Calyx c. 3 mm long, campanulate, pilose with white hairs outside, glabrous inside, tube c. 1.5 mm long, lobes linear, equal to tube. Petals slightly extended from calyx; standard c. 3 mm long, oblong-elliptic with emarginated tip; wing petals shorter than standard and keel petals, c. 2.5 mm long, claw c. 1 mm long; keel petals c. 3 mm long, equal to standard, lamina c. 1 mm long, claw c. 2 mm long. Staminal sheath c. 2 mm long, free filaments minute; vexillary filament c. 2 mm long. Ovary c. 2 mm long, sessile, densely pilose with white hairs; style minute, incurved; stigma capitate. Pods 5–9 x 2.5–3 mm, sessile, spreading star-like in all directions, oblong, deeply grooved dorsally, shortly acuminate, unilocular, covered with appressed or spreading subbasifixed hairs, 4–6-seeded.

Flowering & Fruiting: February–March.

Distribution: INDIA: Punjab; EGYPT; ERITREA; ETHIOPIA; IRAN; IRAQ; OMAN; PAKISTAN; S. ALGERIA; SAUDI ARABIA; SUDAN; YEMEN.

Specimens examined: **Punjab**: Without locality, 1000 ft., *Thomson s. n.* (BM); Ludhiana, *Edgeworth s. n.* (K).

Notes: Based on molecular analysis (Kazempour Osaloo *et al.*, 2003) it was placed under a new genus *Podlechiella* Maassoumi & Kaz. Osaloo. Although, Podlech & Zarre (2013) have followed this treatment, but they have clearly mentioned that morphologically it is more closed to *Astragalus*. Therefore, it has been included here in *Astragalus*. The genus *Podlechiella* has not been accepted in Lock & Schrire (2005).

12. *Astragalus* sect. *Hispiduli*

Podlech, Flora & Vegetatio Mundi 9: 3. 1991; Podlech & Zarre, Tax. Rev. gen. *Astragalus* (Legum.) Old World 1: 118. 2013. –Type species: *Astragalus hispidulus* DC.

Annual herbs. Hairs basifixed, in fruits basifixed to asymmetrically bifurcate, mostly inserted at tubercules. Stipules free or very shortly adnate to petiole. Racemes loose to dense. Calyx tubular or campanulate. Petals unequal in length; wing petals rounded to subacute or emerginate at apex. Style with a ring of minute hairs below stigma. Pods sessile or subsessile, linear, curved, gooved dorsally, completely bilocular.

Key to the Species

1a. Stems with appressed velutinous hairs; standard 5.5–6.5 mm long; pods sessile .... **1. *A. bakaliensis***

1b. Stems with spreading pilose hairs; standard 8–11 mm long; pods with c. 1.5 mm long stipe .... **2. *A. scorpiurus***

1. ***Astragalus bakaliensis*** Bunge, Arbeiten Naturf. Vereins Riga 1: 249. 1847; Ali, Biologia 7: 13. 1962 & in Nasir & Ali, Fl. W. Pakistan 100: 127, f. 17 C. 1977; Kumar & Sane, Legum. South Asia: Checkl. 223. 2003; L.B.Chaudhary & Z.H.Khan, Bull. Bot. Surv. India 47: 53, f. 1. 2005; Podlech in Rech. f., Fl. Iranica 174: t. 18. 1999; Podlech & Zarre, Tax. Rev. gen. *Astragalus* (Legum.) Old World 1: 119. 2013. –Type: Bakali, 28.4.1842, *Lehmann s. n.* (holo P *n. v.*).

Plants prostrate to erect herbs, 3–18 cm long, several branches arising from the base, stems slender, velutinous with appressed white hairs. Stipules c. 1.5 × 1 mm, free, triangular- lanceolate with acute to acuminate tip, glabrous, ciliate along margins. Leaves 10–35 mm long, imparipinnately compound; petiole minute to 8 mm long; rachis 9–25 mm long, hairy as stem; leaflets 4–7 pairs, 3–8 × 1–4 mm, subsessile, close, opposite, oblong or narrowly-ovate, acute at apex, velutinous with spreading, white hairs on both sides. Inflorescence axillary, (1–) 2–4- flowered peduncled raceme, peduncle 15–25 mm long, equal to or longer than subtending leaf, hairy as stem. Bracts minute, persistent, equal to pedicel, ciliate along margins with white or mixed black and white hairs. Flowers 5.5–6.5 mm long, subsessile, purplish-red; pedicel less than 1 mm long, pubescent with white or mixed white and black hairs. Calyx 2–3 mm long, persistent, campanulate, strigose with appressed, mixed black and white hairs outside, tube c. 1.5 mm long, lobes almost equal to tube. Standard 5.5–6.5 × c. 3.5 mm, oblong-elliptic with narrow base and emarginate tip; wing petals 3–4.5 mm long, distinctly shorter than standard and keel petals, lamina c. 3 × 1 mm, oblong with obtuse tip, upper auricle minute, claw c. 1.5 mm long; keel petals c. 6.5 mm long, equal to or slightly shorter than standard, lamina c. 4.5 × 2 mm, upper auricle minute, claw c. 2 mm long. Staminal sheath c. 3.5 mm long, free filaments 2–2.5 mm long ; vexillary filament attached with staminal sheath. Ovary c. 3 mm long, subsessile, strigose with white hairs; stigma minutely penicillate. Pods 8–10 × c. 2 mm, sessile, linear, falcate, turgid, deeply grooved dorsally, acute at tip, partially bilocular with less than 1 mm wide septum, hairy with two types of hairs, longer ones scabrous, unequal, uncinate, white and spreading, shorter ones strigose, appressed and white.

Flowering & Fruiting: April July.

Distribution: INDIA: Jammu & Kashmir; AFGHANISTAN; IRAN; KAZAKHASTAN; PAKISTAN; TADZHIKISTAN; TURKMENISTAN; UZBEKISTAN.

Specimens examined: **Jammu & Kashmir**: Baltistan, Bragicha, 2438 m, 26.5.1928, *Ludlow* 316 (BM); Hushe Nallah, Khare, 2895 m, 18. 6. 1828, *Ludlow* 361 p. p.- specimen marked as A (BM); Shigar, 2566 m, 5.7.1876, *Clarke* 30052B (BM), 30052D & 30052G (CAL). Without Locality: N. W. Himalaya: *Martru* 3243 (CAL).

Notes: It was reported from India by Chaudhary & Khan (2005), but not recorded by Podlech & Zarre (2013).

2. ***Astragalus scorpiurus*** Bunge, Mém. Acad. Imp. Sci. Saint Pétersbourg 11 (16): 5. 1868; Ali, Kew Bull. 1958: 305. 1958; Biologia 7: 16. 1961 & in Nasir & Ali, Fl. W.Pakistan 100: 133. 1977; Sanjappa, Legum. India 94. 1992; Kumar & Sane

Legum. South Asia: Checkl. 240. 2003; Podlech in Rech. f., Fl. Iranica 174: t. 19. 1999; Podlech & Zarre, Tax. Rev. gen. *Astragalus* (Legum.) Old World 1: 126. 2013. –Type: India, Cashmire, *Jacquemont* 14 (lecto P!: photo LWG!; iso: K!).

A. subumbellatus auct. non Klotzsch 1862; Baker in Hook.f., Fl. Brit. India 2: 119. 1876, p. p.

A. punjabicus Sirj., Brittonia 7: 275. 1951. –Type: Pakistan, Rawalpindi, Topi Park, 1700 ft., 24.4.1939, *Stewart* 23605 (holo NY *n. v.*; iso RAW *n. v.*).

A. lasius Blatt., J. Bombay Nat. Hist. Soc. 36: 480. 1933. –Type: Pakistan, N. Waziristan, below Spinwam Fort, 2600 ft., *Blatter* & *Fernandez* 754. (holo n. v.

A. scorpiurus var. *glaber* Ali, Kew Bull. 28: 297. 1973. –Type: Pakistan, Kohat, 21.4.1964, *Afridi s. n.* (holo KUH *n. v.*).

Plate 60

Plants prostrate herbs, several branches arising from the base, 10–35 cm long, stems pilose with white, spreading hairs. Stipules 2.5–3 × 1–2 mm, free, narrowly ovate to ovate, acuminate at tip, glabrescent outside, glabrous inside, ciliate along margins and tip. Leaves 3–4 cm long, imparipinnately compound; petiole 0.5–2 cm long; rachis c. 2 cm long; petiole and rachis hairy as stem; leaflets 7–9 pairs, 4–8 × 1.5–3.5 mm, subsessile, opposite or alternate or subopposite in lower portion, usually ovate, sometimes obovate or oblong, cuneate at base, entire along margins, subacute to obtuse or obtuse-retuse at apex, pilose with white, spreading hairs on both sides. Inflorescence 4–9 cm long, longer than subtending leaf, axillary, peduncled raceme or capitate raceme, 4–10-flowered; peduncle 2–4 cm long, pilose with white, spreading hairs, floral rachis hairy with white, mixed spreading, appressed hairs. Bracts c. 1 mm long, longer than pedicel, ovate-lanceolate with ciliate margins. Flowers 7.5–11 mm long, pink, turning purple; pedicel minute, hairy with white, spreading hairs. Calyx 4–5 mm long, campanulate, glabrous inside, pilose outside, tube 1.5–2 mm long, lobes 2–3 mm long, subequal, shorter on dorsal side, linear lanceolate. Standard 8–11 mm long, ovate, emarginated at apex, claw c. 2 m long; wing petals 7–9 mm long, shorter than standard and keel petals, lamina 5.5–6 × c. 1 mm long, narrowly obovate or oblong with obtuse tip, upper auricle retuse at apex, c. 1 mm long, claw 1.5–2.5 mm long; keel petals 7–10 mm long, subequal to standard, lamina 5–7 × 2–4 mm, curved towards apex, obtuse at apex, claw c. 2 mm long, upper auricle c. 1 mm long. Ovary 3.5–4 mm long, with c. 1.5 mm long stipe, densely hairy, 12–19-ovuld; style 4–5 mm long, incurved in upper part; stigma penicillate. Pods 12–20 mm long, linear, falcate or completely coiled, deeply grooved dorsally, shortly acuminate at apex, with c. 1.5 mm long stipe, almost completely bilocular, 12–24-seeded, hairy, hairs white, mixed long and short, long ones bulbous at base and spreading, short ones very minute and appressed.

Flowering & *Fruiting*: March–April (in plains)–September (in hills).

Habitat: Among riverbeds and grassy places in sandy soils.

Distribution: INDIA: Himachal Pradesh, Jammu & Kshmir, Punjab, Uttarakhand; E. AFGHANISTAN, PAKISTAN.

Specimens examined: **Himachal Pradesh**: Chamba, August 1880, *Ellis s. n.* (CAL). **Jammu & Kashmir**: Sind valley, Gandarbal, 1584 m, 2.6.1940, *Ludlow & Sherriff* 8104 (BM); Sind valley, 1524–1830 m, 23.6.1892, *Duthie* 11441 (CAL: photo LWG). **Punjab:** Faridkot, 19.3.1966, *Nair* 36555 (BSD); Muktesar, 17.3.1966, *Nair* 36528 (BSD). **Uttarakhand:** Dehra Dun, Rispana, 18.4.1965, *Babu* 35206 (BSD); Kumaon, Furkia/ Pindari Moraine, 3500–5000 m, 23–24. 9.1957, *Rao* 4476 (BSD).

Notes: Baker (1876) has treated *Astragalus scorpiurus* Bunge as a synonym of *A. subumbellatus* Klotzsch. However, Ali (1961, 1977) has segregated them based on fruit size and number of seeds in each pod. But he has not recorded these species from India. Podlech & Zarre (2013) have placed them in two different sections and they have also not recorded them from India. Although, Sanjappa (1992) has recorded both species from India, however, the specimens available from India fall under *A. scorpiurus* Bunge.

13. *Astragalus* sect. *Hookeriana*

Podlech, Mitt. Bot. Staatss. Munchen 12: 343. 1976; Podlech & Zarre, Tax. Rev. gen. *Astragalus* (Legum.) Old World 1: 302. 2013. –Type species: *Astragalus acaulis* Baker

Plants nearly acaulescent. Stipules membranous, free from petiole, connate before petiole. Calyx tubular, gibbous at base. Petals yellow. Pods compressed, almost bilocular, without beak, with several transverse fine nerves.

1. ***Astragalus acaulis*** Baker in Hook. f., Fl. Brit. India 2: 132. 1876; Podlech, Mitt. Bot. Staatss. Munchen 12: 345, 348, Abb. 1, 2. 1976; Grierson & Long in Grierson & Long, Fl. Bhutan 1 (3): 716. 1987; Sanjappa, Legum. India 84. 1992; Kumar & Sane, Legum. South Asia: Checkl. 221. 2003; Podlech & Zarre, Tax. Rev. gen. *Astragalus* (Legum.) Old World 1: 303. 2013. –Type: India, Sikkim, Ralung, 16000–17000 ft., *Hooker* (holo K!).

Plants perennial herbs, acaulescent, stems densely tufted at the crown of woody rootstock, up to 3 cm long, glabrous, covered with persistent stipules. Stipules 8–12 × 6–10 mm, persistant, membranous, free from petiole, connate before petiole, encircling the stem, broadly ovate with acute apex (lanceolate in FBI), glabrous or thinly ciliate along margins with white, spreading, long hairs. Leaves 2.5–15 cm long, imparipinnately compoundly compound; rachis pilose with long, white, spreading hairs, subpersistent but not hardened; leaflets 7–12 pairs, 5–10 × 1–2 mm, narrowly ovate or narrowly lanceolate, cuneate or obtuse at base, subacuminate to acuminate at apex, sparsely white hairy only along margins and midrib, otherwise glabrous, hairs long, white, spreading. Inflorescence 1–3-flowered, shortly peduncled head. Bracts 10–13 x c. 2 mm, membranous, linear, ciliate, otherwise glabrous. Flowers yellow, c. 20 mm long; pedicel c. 3 mm long, glabrous. Calyx 10–13 mm long, tubular, tube 7–8 mm long, glabrescent outside, glabrous inside, teeth 3–5 mm long, lanceolate, thinly pilose outside with white, long, spreading hairs, ciliate along margins. Standard c. 20 mm long, lamina c. 14 × 10 mm, broadly elliptic, claw c. 6 mm long; wing petals c. 22 mm long, more or less equal to standard, lamina c. 12 × 3 mm, narrowly oblong, obtuse at apex, claw c. 10 mm long, upper auricle c. 2 × 1.5 mm; keel petals 18–20 mm, slightly shorter than or ± equal to wing, lamina c. 10 × 5 mm, oblong with subobtuse at apex, incurved towards apex, claw c. 10 mm, upper auricle c. 2 × 1.5 mm. Staminal sheath c.18 mm long, free filaments c. 5 mm long; vexillary filament c. 21 mm long. Ovary c.10 mm long, with c. 2 mm long stipe, glabrous; style c. 15 mm long; stigma capitate, glabrous. Pods (after Grierson & Long, 1987 and Podlech & Zarre, 2013) 2.5–5 × 1–1.5 cm, oblong, slightly curved, without beak, almost completely bilocular, glabrous, 10–15-seeded. Seeds c. 4 x 3 mm, dark brown.

Flowering & Fruiting: May–July.

*Distrib*ution INDIA: Sikkim, W. Bengal (upper Darjeeling); BHUTAN; CHINA.

Specimens examined: **Sikkim**: North Sikkim, Gochung opposite, 17. 7. 1986, *Raju* 6069 (BSHC); without locality, *Cave* 5301E (CAL). **W. Bengal**: Upper Darjeeling, Tongolo (Principaute de Kiala), 1893, *Soulie* 711 (CAL).

Notes: It is rare in India. It is chiefly characterized by reduced and glabrous stems, connate, membranous and almost glabrous stipules, lanceolate leaflets with acuminate apex, shortly peduncled 1–3-flowered head Inflorescence, subequal corolla and glabrous pods.

14. *Astragalus* sect. *Hypoglottidei*

DC., Prodr. 2: 281. 1825; Podlech & Zarre, Tax. Rev. gen. *Astragalus* (Legum.) Old World 2: 937. 2013. –Type species: *Astragalus hypoglottis* L.

A. sect. *Hypoglottis* Bunge, Mém. Acad. Imp. Sci. Saint Pétersbourg 11 (16): 46. 1868. –Lectotype species: *Astragalus hypoglottis* sensu Jones, non L.

Perennial herbs, stems short to long, covered with basifixed hairs. Stipules free from petiole, connate opposite the petiole at least in lower leaves. Inflorescence peduncled, globose to shortly cylindrical, dense, many-flowered. Bracteoles absent. Calyx campanulate or tubular, teeth shorter than tube. Petals white, yellow, purple or violet. Pods sessile or shortly stipitate, erect, oblong to ovoid, bilocular (unilocular in *A. gooraiensis* L. B. Chaudhary), generally hairy.

Key to the Species

1a. Calyx 4–5 mm long, campanulate; standard 11–12 mm long; pods oblong-ellipsoid, straight, turgid, unilocular, stipe more or less equal to calyx tube .... 1. ***A. gooraiensis***

1b. Calyx 6.5–8 mm long, tubular; standard 15–20 mm long; pods narrowly oblong to narrowly ellipsoid, slightly falcate, grooved dorsally, bilocular, stipe shorter than calyx tube ....2. ***A. tibetanus***

1. ***Astragalus gooraiensis*** L.B.Chaudhary, Novon 17: 417, f. 1–2. 2007; Podlech & Zarre, Tax. Rev. gen. *Astragalus* (Legum.) Old World 2: 958. 2013. –Type: India Jammu & Kashmir, Goorai, 8000 ft., 21.7.1876, *Clarke* 29345B (holoe BM!; 29345A iso CAL!).

Plants c. 21 cm long, stems sparsely pubescent with minute, white, appressed hairs. Stipules greenish, c. 7 mm long, free from the petiole, leaf opposed, connate at base opposite the petiole, narrowly triangular, acuminate at tip, sparsely pubescent adaxially, glabrous abaxially. Leaves 4–6 cm long, imparipinnately compoundly compound; petiole c. 1 cm long, like the rachis grooved adaxially; rachis c. 5 cm long; leaflets 5–8 pairs, 8–13 × 3–6 mm, oblong, cuneate at base, obtuse at apex, sparsely pubescent with minute, white, appressed hairs below, glabrous above. Inflorescence 6–8 cm long, capitate raceme; peduncle 4.5–5.5 cm long, longer than subtending leaf, hairy as the stem or slightly blackish at apex; raceme c. 1.5 cm long, 6–10-flowered, dense. Bracts greenish, 1.5–2.5 mm long, longer than pedicel, sparsely ciliate with mixed black and white hairs. Pedicel 1–1.5 mm long, pubescent with mixed blackish-brown and white, appressed hairs. Calyx 4–5 mm long, campanulate, oblique at mouth, pubescent with minute, brown or white or mixed brown and white, appressed hairs, more pubescent at teeth, tube c. 3 mm long, upper teeth c. 1 mm long, remote, lower teeth c. 2 mm long, crowded. Standard 11–12 × 6 mm, obovate to elliptic, emarginate at apex, narrowed toward base; wing petals c. 10.5 mm long, slightly shorter than standard, lamina c. 6.5 × 2 mm, oblong or narrowly obovate, slightly curved toward tip, obtuse at apex, auricle straight, oblong rhomboid, subobtuse, c. 1 mm long, claw c. 4 mm long; keel petals c. 9 mm long, shorter than wing petals, lamina c. 4.5 × 3 mm, incurved toward apex, obtuse at apex, auricle minute, triangular, acute, claw c. 4.5 mm long. Staminal sheath c. 6 mm long, obtuse at tip, free filaments 2.5–3 mm long. Ovary c. 2.5 mm long, densely pubescent with white, appressed hairs, with c. 2 mm long stipe; style c. 5 mm long, incurved above; stigma capitate, glabrous. Pods greenish, 11–13 × 4 mm, oblong-ellipsoid, straight, turgid, unilocular, pubescent with minute, blackish-brown, appressed or subappressed hairs, stipe c. 3 mm long, more or less equal to calyx tube.

Flowering & Fruiting: July.

Distribution: INDIA: Jammu & Kashmir. Endemic.

Specimens examined: **Jammu & Kashmir**, Goorai, 2650 m, 21.7.1876, *Clarke* 29345A (CAL), 29345B (BM).

2. ***Astragalus tibetanus*** Benth. ex Bunge, Mém. Acad. Imp. Sci. Saint Pétersbourg 11 (16): 52. 1868 & 12: 85. 1869; Baker in Hook. f., Fl. Brit. India 2: 124. 1876; Gontsch. & Boriss. in Komarov, Fl. U. R. S. S. 12: 258. 1946; Ali, Biologia 7: 43. 1961 & in Nasir & Ali, Fl. W. Pakistan 100: 180. 1977; Sanjappa, Legum. India 95. 1992; Kumar & Sane, Legum. South Asia: Checkl. 243. 2003; Podlech in Fl. Iranica 178: t. 39. 2010; Podlech & Zarre, Tax. Rev. gen. *Astragalus* (Legum.) Old

World 2: 975. 2013. –Type: in Tibeto occ., 9000–14000 ft., *Thomson s. n.* (lecto P!: photo LWG!; iso CAL!: photo LWG!, CGE *n. v.*, E *n. v.*, G *n. v.*, G–BOIS *n. v.*, GOET *n. v.*, L *n. v.*, LE *n. v.*, M *n. v.*, MSB *n. v.*, STU *n. v.*, W *n. v.*).

Plates 17, 61

Plants perennial herbs, decumbent to erect, 7–16 cm long, stems slender, branched, pubescent with appressed white or mixed white and black hairs, the black hairs comparatively longer than white hairs and more on nodal region. Stipules c. 6 × 1.5–2.5 mm, greenish, free from petiole, connate up to less than half of the length or almost up to the tip in the basal portion of stem opposite the petiole, with acuminate tip, ciliate along margins, pilose outside with white or mixed black and white hairs. Leaves 5–10 cm long, imparipinnately compoundly compound; petiole 10–18 mm long; rachis 4–9 cm long, grooved above, pubescent with appressed white or mixed white and black hairs; leaflets 9–14 pairs, 6–9 × 2.5–4.5 mm, subsessile, close, alternate, subopposite or opposite, oblong or oblong-elliptic, cuneate at base, entire along margins, retuse or obtuse at apex, pubescent below with appressed, white hairs, thinly unevenly pubescent to glabrous above. Inflorescence 4–9.5 cm long, axillary, peduncled capitate raceme, more or less equal to or slightly longer than subtending leaf, raceme many-flowered; peduncle 2–8 cm long, pubescent with appressed mixed white and black hairs, pubescent only with black hairs towards tip. Bracts 2–3 (–7) mm long, longer than pedicel, subulate to lanceolate, hairy with mixed black and white hairs, hairs longer on margins and apex. Flowers 15–20 mm long, bluish-white; pedicel c. 1 mm long, pubescent with black hairs. Calyx 6.5–10 mm long, tubular, persistent, tube 5–6.5 mm long, oblique at mouth, pubescent with appressed, black, unequal, intermixed shorter and longer hairs outside, glabrous inside, lobes unequal to almost equal, shorter on dorsal side, densely pubescent with black hairs on both sides, dorsal ones 1–1. 5 (–3.5) mm long, ventral ones c. 1.5 mm long. Standard 15–20 × c. 5 mm, narrowly elliptic, obtuse or minutely retuse at apex; wing petals 13–15 mm long, distinctly shorter or sometimes minutely shorter than standard, lamina c. 7 × 1.5 mm, oblong with obtuse tip, upper auricle and spurs present, claw c. 8 mm long; keel petals 11–13 mm long, shorter than wing petals, lamina c. 4 × 2 mm, slightly curved, upper auricle and spurs present, claw c. 8 mm long. Staminal sheath c. 8 mm long, obtuse at apex, free filaments c. 3–4 mm long; vexillary filament 11 mm long. Ovary c. 3.5 mm long, pubescent with mixed black and white, hairs, with c. 2.5 mm long stipe, convex on dorsal margin; style c. 6 mm long; stigma capitate, glabrous. Pods 11–14 × c. 4 mm, stipitate, narrowly oblong to narrowly ellipsoid, with 1–2 mm long straight to curved beak at apex, slightly falcate (rather recurved), grooved dorsally, pubescent with minute, appressed black hairs, completely bilocular, stipe c. 2 mm long, shorter than the calyx tube.

Flowering & Fruiting: July–September.

Habitat: Growing on mountain slopes along river sides and near agriculture fields among grasses. Altitudes (2300–) 3000–4300 m.

Distribution: INDIA: Jammu & Kashmir; AFGHANISTAN; CHINA; KAZAKHISTAN; KIRGISTAN; MANGOLIA; PAKISTAN; RUSSIA-SIBERIA; TADZHIKISTAN; UZBEKISTAN.

Specimens examined: **Jammu & Kashmir:** Ladakh, Dras, 27.7.1988, *Chowdhery & Uniyal* 85822 (BSD); Dras, 3110 m, 17.7.1976, Collector ? 58700 (BSD); Near Dras town, 3091 m, N34° 25.843′ E75°44.693′, 5.8.2011, *Chaudhary et al.* 263879 (LWG); Ladakh, below Zdongpolas, Shyok River, 3048 m, 5.7.1926, *Ludlow* 395 (BM); Zanskar, Karsha, 3505 m, 18.7.1981, *Southampton University Botanical Expedition* 85 (BM); Suru valley, near Pharma village, 2903 m, N 34°22.378′ E75°58.323′, 4.8.2011, *Chaudhary et al.* 263857 (LWG); Kurgiakh, 4300 m, 25.7.1973, *Bhattacharyya* 52237 (BSD); on way to Pogdan, 6.8.1995, *Balodi* 91084 (BSD); Ladakh, Kangi, 3048–3353 m, July 1905, *Meebold* 1390 (CAL); Buhrawar, 2438 m, 22.9.1876, *Clarke* 31482 B (CAL); Kunalwan, 2286 m, 21.7.1876, *Clarke* 29387 (CAL).

15. *Astragalus* sect. *Hypsophilus*

Bunge, Mém. Acad. Imp. Sci. Saint Pétersbourg 11 (16): 136. 1868; Podlech & Zarre, Tax. Rev. gen. *Astragalus* (Legum.) Old World 3: 2156. 2013. –Type species: *Astragalus nivalis* Kar. & Kir. –Type species: *Astragalus nivalis* Kar. & Kir.

Perennial herbs, stems well developed, hairy with bifurcate hairs. Stipules free from petiole, connate behind the stem (opposite the petiole). Racemes long peduncled, globose, few-flowered. Calyx tubular before anthesis, elongating and becoming inflated like bladder after antheis in fruit, not ruptured by pod. Pods completely enclosed by inflated calyx, shortly stipitate, small, grooved dorsally, compressed laterally, partially bilocular.

1. ***Astragalus nivalis*** Kar. & Kir., Bull. Soc. Imp. Naturalistes Moscou 15: 34. 1842; Bunge, Mém. Acad. Imp. Sci. Saint Pétersbourg 11 (16): 137. 1868 & 12: 234. 1869; Baker in Hook.f., Fl. Brit. India 2: 136. 1876; Gontsch. & Popov in Komarov, Fl. U.R.S.S. 12: 824. 1946; Ali, Biologia 7: 69. 1961 & in Nasir & Ali, Fl. W. Pakistan 100: 198. 1977; Sanjappa, Legum. India 92. 1992; Ohwa *et al.*, J. Jap. Bot. 30: 13, f. 1A. 1994; Kumar & Sane, Legum. South Asia: Checkl. 235. 2003; Podlech in Fl. Iranica 178: t. 373. 2010; Podlech & Zarre, Tax. Rev. gen. *Astragalus* (Legum.) Old World 3: 2158. 2013. –Type: South East of Kazakhstan, Alatau ad fontes fl. Sarchan, *Karelin* & *Kirilov* 272 (lecto MW *n. v.*; iso LE *n. v.*, P!).

Plates 18, 62

Plants perennial herbs, deep rooted, 7–20 cm long, caespitose, prostrate, stems sparsely to moderately strigose, hairs minute, medifixed, completely appressed, white or sometimes black at nodal region. Stipules 3–4 × 2 mm, thick, free from petiole, connate up to the middle in lower portion of the stem and only at base in upper portion of the stem opposite the petiole, completely encircling the stem, lanceolate with acuminate tip, strigose outside with black or white, appressed, medifixed hairs, glabrous inside. Leaves 2.5–4 cm long, imparipinnately compound; petiole 3–15 mm long; rachis c. 20 mm long, strigose, with medifixed, completely appressed, white or sometimes intermixed with black hairs; leaflets 4–7 pairs, 3–5 ×

2–3 mm, subsessile, opposite or sometimes subopposite lower ones, thick, oblong, obovate or broadly obovate, cuneate at base, entire along margins, obtuse or rounded at apex, strigose with white, completely appressed, medifixed hairs. Inflorescence axillary, peduncled capitate raceme or head, many-flowered; peduncle 3.5–4 cm long, longer than subtending leaf, sparsely strigose, with mixed black and white appressed hairs, more hairy towards tip. Bracts 2.5–4 × 1–2 mm, setaceous, hairy, hairs black, intermixed with few white, appressed, mixed medifixed and basifixed. Flowers 18–19 mm long, pinkish-blue; pedicel minute, hairy with mixed black and white hairs. Calyx 10–11 mm long, tubular, tube 9–10 mm long, hairy outside with mixed black and white, mixed basifixed and medifixed hairs, glabrous inside, lobes c. 1 mm long, subequal, densely hairy on both sides, hairs black on outside, white on inside; fruiting calyx inflated, obovoid, membranous. Standard c. 18 mm long, lamina c. 10 × 7 mm, ovate-elliptic, emarginate at apex, claw c. 8 mm long; wing petals c. 16 mm long, lamina c. 6 × 2 mm, oblong, unequally bilobed at tip, upper auricle c. 1 mm long, claw c. 10 mm long; keel petals c. 14 mm long, lamina c. 4 × 2.5 mm, upper auricle minute, spurs present, claw c. 10 mm long. Staminal sheath c. 10 mm long, emarginate at tip, free filaments 1.5–4 mm long. Ovary c. 4 × 1 mm, densely hairy with appressed, mixed light brown and black, basifixed hairs; style c. 8.5 mm long, glabrous, stipe c. 1.5 mm long. Pods 7–9 × c. 3.5 mm, subsessile, included into the calyx, oblong with completely semi-circular upper margin, oblique at base, turgid, partially (half) bilocular, densely pilose with spreading, mixed black and white, basifixed hairs, stipe c. 1 mm long. Seeds 2–2.5 × c. 1.5 mm, reniform, blackish-brown, irregularly compressed and raised.

Flowering & *Fruiting*: July–August.

Habitat: On dry barren hill slopes and near agriculture fields in gravelly or hard soils. Altitudes 4000–5500 m.

Distribution: INDIA: Himachal Pradesh, Jammu & Kashmir; AFGHANISTAN; CHINA; KAZAKHSTAN; KIRGISTAN; PAKISTAN; TADZHIKISTAN.

Specimens examined: **Himachal Pradesh**: Lahul-Spiti, 58 km before Pang from Sarchu, 4200 m, 5.8.2003, *Chaudhary & Khan* 225034 (LWG); Sarchu, 30.6.1941, *Bor* 15105 (DD); 3 km before Sarchu, 4332 m, N 32° 54.014' E 77° 33.217', 30.7.2011, *Chaudhary et al.* 263821 (LWG); Between Sarchu and Pang, 8 km before Pang from Sarchu, 4. 8. 2014, Chaudhary *et al.* 259370 (LWG); Takcha, 13500 ft., 21.7.1958, *Kaushik* 1086 (DD); Lossar, 9.8.1994, *Murti & Singh* 88209 (BSD); Keylang, 4572 m, 20.7.1941, *Bor* 16477 (DD); Lossar, 4100 m, 26.7.1972, *Bhattacharyya* 48891 (BSD, CAL); Lossar, near PWD guesthouse, 4000 m, 4.8.2002, *Chaudhary & Khan* 206762 (LWG); Lossar, 4079 m, N32° 25.848′ E77°43.055′ , 2.8.2006, *Chaudhary et al.* 229519 (LWG); before 2 km of Lossar from Kunjam pass, 4101 m, N 32°26.858′ E 76°43.455′, 1.8.2006, *Chaudhary et al.* 229510 (LWG); before 1 km of Lossar from Kunjam pass, 4106 m, N 32°26.590′ E 76° 44.573′ , 1.8.2006, *Chaudhary et al.* 229507 (LWG); 1 km before Lossar, 4000 m, 4.8.2002, *Chaudhary & Khan* 206758 (LWG); Thumla, 4200 m, 1.8.1972, *Bhattacharyya* 49200 (BSD); Khurik, 3800 m, 6.8.1972, *Bhattacharyya* 49371 (BSD); Pangi, 3353–5029 m, 1879, *Heyde s. n.* (CAL). **Jammu & Kashmir**: Ladakh, Fotula, 4500 m, 25.8.1986, *Aswal & Goel* 16365 (CDRI); Fotula, 3990 m, 25.7.1876,

Wadhwa 58964 (BSD); Lingii, 4800 m, 29.7.1973, *Bhattacharyya* 52313 (BSD); Lingti, 4500 m, 31.7.1973, *Bhattacharyya* 52327 (BSD); La Chang La, between Sarchu and Pang, 4702 m, N 33° 01.702′ E 77° 37.534′, 31.7.2011, *Chaudhary et al.* 263824 (LWG); Pang-Slopes above Military Post, 4570 m, *Goraya et al.* 1015 (BSD); Ladakh, Kangi, 3048–3353 m, July 1905, *Meebold* 1387 (CAL); Ladakh, 5029 m, 31.7.1928, *Ludlow* 462 (BM); Ladakh, Chong Kumdam, Shegoh village, 4815 m, 26.7.1929, *Ludlow* 566 (BM).

16. *Astragalus* sect. *Irinaea*

Boriss. in Komarov, Fl. U.R.S.S. 12: 880. 1946; Podlech & Zarre, Tax. Rev. gen. *Astragalus* (Legum.) Old World 2: 926. 2013. –Type species: *Astragalus saxifractor* Rech. f. & Gilli.

Perennial herbs, with well developed stems, covered with basifixed to bifurcate hairs. Stipules shortly adnate to petiole or free. Racemes peduncled, 5–10-flowered. Calyx campanulate. Petals almost of the same length. Stigma glabrous or rarely penicillate. Pods oblong to ellipsoid, generally upcurved, grooved dorsally, unilocular or bilocular.

1. ***Astragalus amherstianus*** Royle ex Benth. in Royle, Illustr. Bot. Himal. Mount. 199. 1835; Baker in Hook.f., Fl. Brit. India 2: 119. 1876; Ali, Biologia 7:12.1961; Bot. Notiser 118: 88. 1965 & in Nasir & Ali, Fl. W. Pakistan 100: 127, f. 17 B–B1, 18 A–G. 1977; Sanjappa, Legum. India 84. 1992; Kumar & Sane, Legum. South Asia: Checkl. 222. 2003; Podlech in Fl. Iranica 178: t. 3. 2010; Podlech & Zarre, Tax. Rev. gen. *Astragalus* (Legum.) Old World 2: 927. 2013. –Type: India: Simla, *Amherst s. n.* (holo K!).

A. lowarensis Ali, Kew Bull. 28: 299. 1973. –Type: Pakistan, Chitral, Ziarat, Lowarai Pass, 6500 ft., 29.4.1958, *Stainton* 2318 (holo BM *n. v.*; iso E *n. v.*, LE *n. v.*).

Plates 19, 63

Plants perennial herbs, caespitose, 12–45 cm long, stems procumbent to ascending, slender, strigose with white, appressed, forwardly oriented hairs. Stipules 2–3.5 × 1–1.5 mm, free from petiole, spreading, triangular-ovate or ovate, acute to acuminate at tip, strigose outside, glabrous inside, hairs similar to stem. Leaves 3–4 cm long, imparipinnately compound; petiole 3–7 mm long, strigose like stem; rachis 2.7–3 cm long, channelled above, strigose like stem; leaflets 6–9 pairs, 3–9 × c. 3 mm, subsessile, opposite, subopposite or alternate lower ones, oblong, narrowly obovate, ovate or obovate (pariticularly upper ones), cuneate at base, entire along margins, retuse, subtruncate, obtuse or acute at apex, velutinous on both sides with white, long, spreading hairs (in herbarium specimens hairs appear appressed), densely hairy in young stage. Inflorescence 25–35 mm long, almost equal to subtending leaf, axillary, peduncled raceme, 5–10-flowered; peduncle 13–25 mm long, strigose like stem; rachis 8–13 mm long, strigose like stem. Bracts c. 1 mm long, setaceous, hairy. Flowers 4.5–5 mm long, pale creamy or white with purple tip; pedicel c. 1 mm long, hairy. Calyx 4–4.5 mm long, persistent, campanulate, 5-lobed, velutinous with white, spreading hairs outside, glabrous inside, tube c. 1.5 mm long, lobes almost double the tube, 2.5–3 mm long, subequal, slightly shorter upper ones, filiform. Standard c. 5 mm long, lamina c. 4 × 3 mm, ovate, emarginate at apex, claw c. 1 mm long; wing petals c. 4.5 mm long slightly shorter than standard and keel petals, lamina c. 2.5 × 0.7 mm, narrowly oblong, slightly curved, obtuse at apex, upper auricle and spurs present, claw c. 2 mm long; keel petals c. 4.7 mm long, slightly shorter than standard and longer than wing petals, lamina c. 2.7 × 1.8 mm, incurved on lower margins, obovate with subobtuse tip, upper auricle and spurs present, claw c. 2 mm long. Staminal sheath c. 4 mm long, obtuse at tip, free filaments 1–1.5 mm long, alternately shorter and longer; vexillary filaments c. 4.5 mm long. Ovary c. 2.5 mm long, subsessile, linear, densely velutinous with white hairs; style c. 2 mm long, incurved in upper part, glabrous; stigma capitate, glabrous. Pods 8–12 × c. 3 mm, sessile, narrowly oblong, turgid, semilunar, pointed at tip, unilocular, grooved dorsally, dehiscent, c. 7-seeded, pilose with white, spreading hairs. Seeds c. 2 × 1.5 mm, reniform, brown, turgid, smooth, glabrous.

Flowering & *Fruiting*: July–August.

Habitat: On shaddy or open moutain slopes in exposed sandy soil or on sandy river beds. Altitudes 2000–4200 m.

Distribution: INDIA: Himachal Pradesh, Jammu & Kashmir, Uttarakhand; AFGHANISTAN; NEPAL? (not recorded by Podlech & Zarre, 2013); PAKISTAN.

Specimens examined: **Himachal Pradesh**: Lahul: Jispa, 3500 m, 19.8.1982, *Aswal* 11016 (CDRI); Kirting, 3100 m, 6.9.1981, *Aswal* 10856 (CDRI); Jispa, Bhoga valley, 3300 m, 14.8.1970, *Bhattacharyya* 40788 (BSD); Jispa, 3200 m, 9.7.1973, *Bhattacharyya* 51902 (BSD); Tandi, 2900 m, 5.8.1971, *Bhattacharyya* 45128 (BSD); Tandi, 3000 m, 29.7.1979, *Aswal* 10468 (CDRI); Keylang, Gondla, 5.6.1941, *Bor* 10300 (DD); Pattan valley, Thirot, 2743 m, 14.6.1958, *Sethi* & *Negi* 339 (DD); Lat, 3048 m, 16.6.1941, *Bor* 9729 (DD); Chamba, Bharmour, 2150 m, 27.7.63, collector? 28322 (BSD); On the way to Udaypur, Lote, , 2931 m, N 32° 36.652′ E 76° 55.349′, 30.7. 2006, *Chaudhary et al.* 229472, (LWG); Near Trilokinath Temple, 3845 m, N 32° 47.265′ E 77°17.099′, 30.7.2006, *Chaudhary et al.* 229477, (LWG). **Jammu & Kashmir:** Himalya Bor. Occ. Region temp. 1830–2134 m, *Thomson s. n.* (BM). **Uttarakhand**: Chamoli, on way to Sumna, *Hajra* 87590 (BSD); way to Dunagiri, 2600 m, 16.8.1974, *Naithani* 54044 (BSD); Uttarkshai, Harsil, N 31°02.33′ E 78°48.580′, 2502 m, 2.8.2006, *Chaudhary et al.* 229556 (LWG); Uttarkashi, way to Nelang, July 1988, *Naithani* 66497 (BSD); Gangotri, 3657–3962 m, Oct., 1881, *Duthie* 1494 (CAL); Between Banas & Kasali bed of Jumna, 2438–2743 m, Oct 1877, *Duthie* 134 (CAL).

Notes: In the above specimens the stigma are glabrous and calyx teeth are distinctly twice longer than tube. However, Baker (1876) has observed penicillate stigma and calyx teeth equal to tube in this species. Ali (1977) has also noticed penicillate stigma in this species.

17. *Astragalus* sect. *Komaroviella*

Gontsch. in Komarov, Fl. U.R.S.S. 12: 876. 1946; Podlech & Zarre, Tax. Rev. gen. *Astragalus* (Legum.) Old World 2: 904. 2013. –Lectotype species: *Astragalus alpinus* L.

Perennial herbs, subcaulescent to caulescent, hairy with basifixed hairs. Stipules free from petiole, connate all or only lower ones or rarely all free. Racemes peduncled, ovoid to subumbellate, loose. Bracteoles mostly absent. Calyx campanulate. Petals violet, wing petals shorter or slightly longer than keel petals. Pods pendulous, incompletely to completely bilocular, hairy or rarely glabrous, stipe equal to or slightly longer than calyx.

Key to the Species

1a. Stipules ciliate, otherwise galbrous; leaflets glabrous on upper surface, hairy on lower surface; pods glabrous .... **1. *A. alpinus***

1b. Stipules hairy outside, glabrous inside; leaflets hairy on both sides; pods hairy 2

2a. Calyx campanulate-tubular, lobes longer than tube; pod stipe shorter than calyx tube .... **5. *A. oxyodon***

2b. Calyx campanulate, lobes equal to or shorter than tube; pod stipe equal to or longer than calyx tube 3

3a. Stupules free or shortly adnate to the petiole, connate infront of the petiole as well as opposite the petiole .... **2. *A confertus***

3b. Stupules free from the petiole, connate only opposite the petiole 4

4a. Stamens closed monadelphous, rarely diadelphous; pod stipe longer than calyx .... **4. *A. himalayanus***

4b. Stamens diadelphous (9+1); pod stipe almost equal to calyx .... **3. *A. himachalensis***

1. *Astragalus alpinus* L., Sp. Pl. 760. 1753; Baker in Hook. f., Fl. Brit. India 2: 123. 1876; L.B. Chaudhary & Z.H. Khan, Phytotaxon. 4: 57, f. 1 & 4 A–F. 2004; Xu & Podlech in Wu *et al.*, Fl. China 10: 359, f. 385, 8-14. 2010; 2010; Podlech & Zarre, Tax. Rev. gen. *Astragalus* (Legum.) Old World 2: 907. 2013. –Type: Switzerland, in pascuis graminosis montis "Joch-Alpe" prope Churwalden, July 1863, *Brügger s. n.* (epi BM *n. v.*). Scheuchzer, Uresif. Helv.: f. 7 (inter pp. 508 & 509). 1723 (lecto: icono).

<u>Plants</u> perennial spreading herbs, c. 10 cm long, stems slender with elongated internodes, glabrous in lower portion, pilose with white, appressed hairs in upper portion. <u>Stipules</u> c. 3 × 1.5 mm, free from petiole, connate up to more than half of the length opposite the petiole, ovate, acute at apex, ciliate with white hairs along margins otherwise glabrous. <u>Leaves</u> 3.5–5 cm long, imparipinnately compound;

petiole c. 11 mm long; rachis thinly pilose with white, appressed hairs; leaflets (6–7–) 8–10 pairs, 3–6.5 × 1–2.5 mm, opposite, subopposite or alternate, subsessile, thick, ovate, oblong-ovate or oblong, acute at apex, pilose only along midrib region on lower surface, ciliate along margins and extra marginal portion on upper surface with white appressed hairs. Inflorescence c.10 mm long, shorter than subtending leaf, axillary, peduncled raceme, 2–8-flowered; peduncle c. 8 mm long, elongating in fruits, pilose with appressed, white hairs. Bracts c.1.5 mm long, persistent, equal to pedicel, ciliate with white hairs. Flowers c. 9 mm long, violet; pedicel c. 1.5 mm long, thinly pilose with appressed white or intermixed brown hairs. Calyx c. 4 mm long, campanulate, pilose with appressed, mixed white and brownish hairs, tube c. 2 mm long, lobes ± equal to tube, linear-lanceolate. Standard c. 9 mm long; wing petals more or less equal to standard; keel petals c. 6 mm long, distinctly shorter than standard and wing petals. Pods c. 18 × 4, oblong, straight, pointed at both ends, turgid, grooved dorsally, glabrous (hairy in Podlech & Zarre, 2013), stipitate, unilocular, stipe slightly shorter than calyx tube.

Flowering & Fruiting: August.

Distribution: INDIA: Jammu & Kashmir; CHINA; PAKISTAN; NORTH TEMPERATE ZONE.

Specimens examined: **Jammu & Kashmir**: Ladakh, Chushul, 5500 m, 24.8.1975, *Vishwanathan* 55012 (BSD).

2. *Astragalus confertus* Benth. ex Bunge, Mém. Acad. Imp. Sci. Saint Pétersbourg 11 (16): 23. 1868 & 15 (1): 27. 1869; Baker in Hook. f., Fl. Brit. India 2: 123. 1876; Ali, Biologia 7: 42. 1961 & in Nasir & Ali, Fl. W. Pakistan 100: 160. 1977; Sanjappa, Legum. India 86. 1992; Kumar & Sane, Legum. South Asia: Checkl. 226. 2003; Podlech & Zarre, Tax. Rev. gen. *Astragalus* (Legum.) Old World 2: 914. 2013. –Type: Tibet, Pangong, 17500 ft., 1852, *Strachey* 15 (lecto K!).

Plate 64

Plants perennial herbs with woody rhizome, prostrate, 2–10 cm long, densely caespitose, stems many, very dwarf, thinly pubescent with white, appressed hairs. Stipules minute, lanceolate, free from or shortly adnate to the petiole, connate infront of petiole as well as opposite the petiole, sparsely to densely hairy, glabrescent to glabrous at maturity. Leaves 10–25 mm long, imparipinnately compound; petiole 5-8 mm long, subappressed to ascending white hairy; leaflets 5–7 pairs, 1.5–3 × c. 1 mm, close, oblong, subacute to obtuse at apex, densely appressed white hairy on both sides. Inflorescence axillary, peduncled raceme, 1–8-flowered; peduncle 2.5–5 cm long, longer than subtending leaf, pubescent with white, appressed hairs in lower portion and black hairs in upper portion. Bracts c. 2 mm long, ovate to narrowly triangular, prominently black ciliate. Flowers violet, c. 13 mm long; pedicel c. 1 mm long, densely black hairy. Calyx 4–5 mm long, campanulate, pubescent with mostly black, appressed hairs, tube 2.5–3 mm long, slightly oblique at mouth, lobes 1.5–2 mm long, shorter than tube, unequal, narrowly triangular, densely hairy on inner side. Standard 10–12 mm long; broadly elliptic to suborbicular, emarginated at apex, contracted into a short claw at base; wing petals 8–10 mm long, oblong, obtuse to roundea at apex, auricle c. 0.5–1 mm long, claw 3–4.5 mm long; keel petals 9–11 mm long, shorter than standard, obovate, inwardly curved in upper portion, obtuse at apex, auricle miniute, claw 2–4 mm long. Staminal tube more or less truncate at apex. Ovary white hairy, with 2–2.5 mm long stipe; style glabrous. Pods unilocular, few-seeded, appressed white pubescent, stipe c. 3 mm long.

Flowering & *Fruiting*: July–September.

Habitat: Altitudes 4200–6000 m.

Distribution: INDIA: Himachal Pradesh , Jammu & Kashmir; CHINA; NEPAL.

Specimens examined: **Himachal Pradesh:** Pangi, *Dovie* 87 (CAL). **Jammu & Kashmir**: Rupchu, 4572–5486 m, June 1865, *Stoliczka s. n.* (CAL); Leh-Ladakh, Tanglang La, 5347 m, 31.7.2011, *Chaudhary et al.* 263827 (LWG).

Notes: It is much closed to *A. strictus* Benth. Both speices can be differentiated by habit of the plants and the length of the keel petals. In *A. confertus* the plants are very dwarf with reduced stems and keel petals are subequal to wing petals and shorter than standard, while in *A. strictus* the stems are well developed (up to c. 30 cm long) and keel petals are distinctly shorter than wing petals.

3. ***Astragalus himachalensis*** L.B.Chaudhary & T. S. Rana, Bot J. Linn Soc. 154: 31, f. 3. 2007. –Type: India, Himachal Pradesh, Lahul-Spiti, Keylang, Jispa, 14.9.2001, *Chaudhary* 220659 (holo LWG!; iso CAL!).

Plates 20, 65

Plants perennial herbs, prostrate, densely caespitose, stems 5–15 cm long, slender, covered with strigose, white, appressed, hairs. Stipules 2–3 (–5) × c. 1 mm, spreading, free from petiole, leaf opposed, connate at base and encircling the stem opposite the petiole, triangular-lanceolate, acuminate at tip, ciliate along margins, strigose outside, glabrous inside. Leaves 1.1–3.5 (–4) cm long, imparipinnately compound; petiole 0–6 mm long, strigose; leaflets 8–11 pairs, 2–4 (–6–9) × 1–2.5 (–3.5) mm, close, opposite, subopposite or alternate, ovate, oblong or oblong-elliptic, cuneate or obtuse at base, entire along margins, acute, subobtuse or obtuse at apex, pilose with white, spreading hairs on both sides. Inflorescence peduncled capitate raceme; peduncle 0.7–2.5 (–4) cm long, strigose with appressed, white or mixed black and white hairs; raceme 5–8-flowered, close. Bracts 1–2 mm long, persistent, linear-setaceous, entire or with a few teeth in basal portion and ciliate along margins, pilose. Flowers 8–9 mm long, purple or bluish-purple; pedicel c. 1 mm long, strigose. Calyx 2.5–3 (–4) mm long, campanulate, tube c. 2 mm long, slightly oblique at mouth,strigose outside, glabrous inside, lobes 0.5–1 (–2) mm long, unequal, shorter on dorsal side, strigose on both sides. Standard 8–9 mm long, lamina 6–6.5 × 5–6 mm, broadly ovate-elliptic, emarginate at apex, claw 2–2.5 mm long; wing petals 7.5–8 mm long, lamina 4.5–5.2 × c. 1.5 mm, oblong-obovate or oblong-elliptic, obtuse or subacute at tip, claw 2.5–3 mm long; keel petals 8–9 mm long, lamina c. 5 × 3 mm, oblong, incurved towards and obtuse at apex, claw 3– 4 mm long. Staminal sheath 6–7 mm long, obtuse at apex, free filaments 2–3 mm long; vexillary filament c. 7 mm long. Ovary 2–2.5 mm long, densely hairy, c. 4-ovuled, stipe 2–2.5 mm long, shorter than calyx; style 4–6 mm long, persistent, curved in upper part, glabrous; stigma capitate, glabrous. Pods 5–6 (–8) × 2.5–3.5 mm, oblong or oblng-ellipsoid, turgid, abruptly narrowed at tip, obtuse or occasionally cuneate and curved at base, stipitate, strigose with appressed, white or black or mixed black and white hairs, faintly veined, unilocular, 2–4-seeded, stipe 2–3 (–4) mm long, almost equal to or slightly shorter than calyx. Seeds c. 2.8 × 1.5 mm, reniform, blackish-brown, turgid, smooth, glabrous.

Flowering & Fruiting: July–September.

Distribution: INDIA: Himachal Pradesh. 3300–4400 m. Endemic.

Specimens examined: **Himachal Pradesh**: Lahul-Spiti: Baralach La, 4400 m, 4.8.2003, *Chaudhary & Khan* 225026 (LWG); Near Pitsio, 3800 m, 4.8.2003, *Chaudhary & Khan* 225022 (LWG); Near Pitsio, 3785 m, N32° 45.567′ E77° 15.602′, 31.07.2011, *Chaudhary et al.* 263812 (LWG); Jispa, 3400 m, 4.8.2003, *Chaudhary & Khan* 225015A (LWG); Jispa, 3278 m, N32° 37.960′ E77°10.442′, 29.7.2006, *Chaudhary et al.* 229434 (LWG); Chotadhara, 3700 m, 3.8.2002, *Chaudhary & Khan* 206751 (LWG); Chotadhara, 18.7.2008, *Chaudhary et al.* 249610 (LWG); Chotadhara, 3475 m, N32° 18.035′ E77°27.704′, 31.7.2006, *Chaudhary et al.* 229487, 229491 (LWG); Between Chotadhara and Batal, 3512 m, N32° 17.654′ E77°31.297′, 31.7.2006, *Chaudhary et al.* 229505 (LWG); Between Chatru & Chotadhara, 3027 m, N32° 20.079′ E77°20.413′, 31.7.2006, *Chaudhary et al.* 229495 (LWG); 5 km before Chattru towards Chotadhara, 3395 m, N 32° 18.608′ E 77° 23.322′, 5.8.2012, *Chaudhary & Bajpai* 259323 (LWG); Near Chatru, 3034 m, N32°21.654′ E77°18.247′, 31.7.2006, *Chaudhary et al.* 229499 (LWG); Near Chatru, 3300 m, 3.8.2002, *Chaudhary & Khan* 206747 LWG); Chatru, 3300 m, 3.8.2002, *Chaudhary & Khan* 206748 (LWG); Before 2 km Chatru, 18.7.2008, *Chaudhary et al.* 249609 (LWG).

Notes: Podlech & Zarre (2013) have synonymised this species under *A. himalayanus* Klotzsch perhaps by considering only stamen character. While describing *A. himachalensis*, it was thought that both also differ in stamen character (diadelphous in *A. himachalensis* and monadelphous in *A. himalayanus),* but it has been later on observed that in *A. himalayanus* both conditions are found. In spite of this, both species differ in many other morphological characters in addition to molecular differences (Chaudhary *et al.*, 2007).

4. ***Astragalus himalayanus*** Klotzsch in Klotzsch & Garcke, Bot. Ergebn. Reise Waldemar 160, t. 4. 1862; Baker in Hook. f., Fl. Brit. India 2: 126. 1876; Ali, Biologia 7: 40. 1961 & in Nasir & Ali, Fl.W. Pakistan 100: 179. 1977; Chater in Hara & Williams, Enum. Fl. Pl. Nepal 2: 107. 1979; Wenninger, Mitt. Bot. Staatssamml. Munchen 30: 123, f. 11 e, f. 24 m-o. 1991; Sanjappa, Legum. India 89. 1992; Kumar & Sane, Legum. South Asia: Checkl. 230. 2003; Podlech & Zarre, Tax. Rev. gen. *Astragalus* (Legum.) Old World 2: 917. 2013, p. p. (excluding *A. himachalensis* L.B.Chaudhary & T.S. Rana). –Type: The table 4 in Bot. Erg. Waldemar Reise. 1862 (neo).

A. maddenianus Benth. ex Baker in Hook. f., Fl. Brit. India 2: 127. 1876; Ali, Biologia 7:39. 1961 & in Nasir & Ali, Fl. W. Pakistan 100: 179. 1977; Sanjappa, Legum. India 90. 1992. –Type: India, Himalaya, Badrinath, 10000–11000 ft., *Madden s. n.* (lecto K!).

Plates 21, 66

Plants perennial herbs, 20–60 cm long, procumbent to suberect, sometimes growing like climber, branched, stems elongated, slender, pubescent (apical portion more hairy), hairs white, forwardly oriented, completely appressed or occasionally spreading. Stipules 3.5–4.5 × 0.8–1 mm, persistent, free from petiole, connate in lower portion opposite the petiole, spreading free part triangular-lanceolate, acuminate at tip, ciliate along margins with long, spreading, white hairs, glabrate to sparsely hairy outside, glabrous inside. Leaves 3.5–4 cm long, imparipinnately compound; petiole 2.5–4 (–7) mm long, hairy as stem; rachis 2.8– 3.2 cm long, channeled above, hairy as stem; leaflets 8–11 pairs, 4.5–10 × 1.8–5 mm, alternate, subopposite or opposite, minutely petiolulate, oblong, narrowly obovate or ovate, cuneate at base, entire along margins, truncate, retuse, obtuse or rounded at apex, pilose with white spreading hairs both sides. Inflorescence axillary, peduncled condensed raceme, 3.5–7 cm long, smaller than or sometimes equal to subtending leaf, 3–10-flowered; peduncle 1.8–4 cm long, pubescent, hairs white or mixed black and white, appressed. Bracts 0.5–3 × 0.5–1 mm, persistent, quite variable, sometimes with reddish tinge, shorter or equal to pedicel, linear to linear-ovate, ciliate along margins with white or mixed black and white long, spreading hairs, intermixed with minute gland tipped hairs, sometimes lobed towards basal portion, acute or acuminate at tip, glabrate outside, glabrous inside. Flowers purple to bluish-purple, 9–14 mm long; pedicel 1–2 mm long, variable in length, densely pubescent with black or mixed black and white, appressed to subspreading hairs. Calyx 2.5–4 mm long, persistent, campanulate, 5-lobed, tube 2–3 mm long, pilose with completely black or mixed black & white appressed hairs outside, glabrous inside, sometimes tinged with red colour on dorsal side, lobes 0.5–1 (–2) mm long, shorter than or sometimes equall to tube, shorter on dorsal side, triangular, densely villous on both sides with black or mixed black & white hairs. Standard 9–12 mm long, deflexed, lamina 6–8 × c. 6 mm, ovate to broadly ovate, emarginate at apex, claw 3–4 mm long; wing petals 7–10.5 mm, distinctly shorter than standard and keel petals, lamina c. 4 × 1.5–2 mm, oblong-obovate, obtuse at tip, without spurs, upper auricle c. 1.5 mm long with truncated tip, claw 3–3.5 mm long; keel petals 9–12.5 mm long, subequal to standard, lamina 5–8.5 × 2.8–3 mm, oblong, slightly curved at apex, upper auricle c. 1 mm long, spurs faintly developed, claw c. 4 mm long. Stamens usually monadelphous (closed) or sometimes diadelphous, 8.5–11.5 mm long; staminal sheath 7–9 mm long, free filaments 2.5–3.5 mm long. Ovary c. 3 mm long, linear, appressed hairy with white hairs, 4–6-ovuled, with c. 5 mm long stipe; style c. 4 mm long, curved towards apex; stigma capitate, glabrous. Pods 12–18 × 3.5–4 mm, narrowly oblong or narrowly ellipsoid, stipitate, generally grooved dorsally, turgid, unilocular, with membranous valves, reticulately veined, tapering at both ends, pubescent with spreading or appressed and forwardly oriented, mixed black and white hairs, sometimes densely covered only with black hairs, 4–6-seeded, stipe 5–6 mm long, longer than calyx. Seeds 2–2.5 × 1.5–1.8 mm, reniform, turgid, brownish-black, smooth, glabrous.

Flowering & Fruiting: July–September.

Habitat: Commonly growing on mountain slopes, flat lands or near agricultural fields in moist and shaddy places among grasses and herbs. Altitudes 1700–4400 m.

Distribution: INDIA: Himachal Pradesh, Jammu & Kashmir, Uttarakhand; PAKISTAN; NEPAL. Podlech & Zatrre (2013) have not recorded it from Nepal.

Specimens examined: **Himachal Pradesh**: Lahul, Chandra valley, 3657 m, 19.8.1952, *Graff* 13 (BM); Lahul-Spiti, Khoksar, 13.9.2001, *Chaudhary* 220638 (LWG); Khoksar, 16.9.2001, *Chaudhary* 220676 (LWG); Khoksar, 3400 m, 2.9.1961, *Nair* 16476 (BSD); Khoksar, 3150 m, 2.8.2002, *Chaudhary & Khan* 206739 (LWG); On way to Khoksar towards Chhatru, 3300 m, 3.8.2002, *Chaudhary & Khan* 206744 (LWG); Khoksor, 3500 m, 11.7.1977. *Aswal* 6593; Khoksar, 3200 m, 11.7.1977, *Aswal* 6560 (CDRI); Khoksar, 3200 m, 12.7.1977, *Aswal* 6593 (CDRI); Khoksar, 3200 m, 14.7.1979, *Aswal* 10203 (CDRI); Khoksar, 3200 m, 15.9.1978, *Aswal* 6988 (CDRI); Khoksar, 3400 m, 2.9.1961, *Nair* 16498 & 16498b (BSD); Khoksar, 3375 m, 15.9.1961, *Nair* 17007 (BSD); Khoksar, 3400 m, 2.9.1961, *Nair* 16498 (BSD); Khoksar, 3200 m, 2.8.1971, *Bhattacharyya* 44954 (BSD); Khoksar, 3200 m, 18.7.1972, *Bhattacharyya* 48689 (BSD); Khoksar, 3200 m, 4.8.1970, *Bhattacharyya* 40525 (BSD); Khoksar , 3136 m, N 32° 24.755′ E 77°14.180′, 28.7.2006, *Chaudhary et al.* 229415 (LWG); Khoksar , 3134 m, N 32° 24.809 ′E 77°14.225′, 28.7.2006, *Chaudhary et al.* 229420 (LWG); Khoksar-Sissu, 3400 m, 26.6.1958, *Rau* 5878 (BSD); near Sissu, 13.9.2001, *Chaudhary* 220643 (LWG); Between Khoksar and Sissu , 3055 m, N 32° 27.149 ′E 77° 09.074′, 28.7.2006, *Chaudhary et al.* 229426 (LWG); Keylang, 3048 m, 20.6.1958, *Sethi & Negi* 415 (DD); Keylang, near Gamur village, 14.9.2001, *Chaudhary* 220654 (LWG); Keylang, 3048 m, 20.6.1958, *Sethi & Negi* 415 (DD); Before Keylong and Jispa , 3337 m, N 32° 35.374′ E 77° 07.640′, 29.7.2006, *Chaudhary et al.* 229429(LWG); Chamba, Bindrabani, 2500 m, 17.7.1964, *Nair* 32502 (BSD); Trilokinath, U. Chenab, 3000 m, 21.8.1971, *Bhattacharyya* 45903 (BSD); Above Marhi, 3814 m, N 32° 24.427′ E 77°14.806′, 28.7.2006, *Chaudhary et al.* 229413 (LWG); Above Marhi near Rohtang, 3450 m, 2.8.2002, *Chaudhary &. Khan* 206734 (LWG); Before Rohatang , 3814 m, N 32° 24.427′ E 77° 14.806′, 28.7.2006, *Chaudhary et al.* 229410 (LWG); Below Rohatang pass, 3814 m, N 32° 24.427 ′E 77°14.806′, 28.7.2006, *Chaudhary et al.* 229414 ((LWG); Rohtang (6 km before from Manali side), 3.8.2003, *Chaudhary & Khan* 225002 (LWG); Between Marhi to Rohtang, 3500 m, N 32° 21.346′ E 77°13.533′, 17.7.2008, *Chaudhary et al.* 249601 (LWG); 4 km away from Khoksar towards Sissu, 3200 m, 3.8.2003, *Chaudhary & Khan* 225006 (LWG); between Khoksar & Rohtang, 4475 m, 16.9.1961, *Nair* 17064 (BSD); Keylang, near P.W.D. resthouse, 15.9.2001, *Chaudhary* 220674 (LWG); Lahul, Jispa, 3400 m, 17.9.1978, *Aswal* 9365 (CDRI); Jispa, 3330 m, 17.9.1978, *Aswal* 9363 (CDRI); Jispa, 3400 m, 4.8.2003, *Chaudhary & Khan* 225015 (LWG); Jispa, 3400 m, 4.8.2003, *Chaudhary & Khan* 225016 (LWG); between Tandi & Gondla, 3100 m, 3.8.2003, *Chaudhary & Khan* 225011 (LWG); Tandi-Gondla, 3200 m, 23.9.1978, *Aswal* 10067 (CDRI); Gondla, 3400 m, 6.7.1973, *Bhattacharyya* 51876 (BSD); Gondla-Sissoo, 7.8.1994, *Murti & Singh* 88138

(BSD); Jispa, 3278 m, N 32°37.960'E 77°10.442', 29.7.2006, *Chaudhary et al.* 229451 (LWG); Gondla , 2996 m, N 32° 30.842'E 77° 58.987', 30.7.2006, *Chaudhary et al.* 229483 (LWG); Gramphu , 2996 m, N 32°30.842'E 76° 58.987', 31.7.2006, *Chaudhary et al.* 229497 (LWG); Gramphoo-Dorni, 3300 m, 30.9.1978, *Aswal* 10151 (CDRI); Lahul, near Pitscio, 3800 m, 4.8.2003, *Chaudhary & Khan* 225022 (LWG); Chhatru, 3375 m, 15.9.1961, *Nair* 17007 (BSD); Rakcham, 12500 ft., 12.8.1973, *Janardhanan* 52654 (BSD); Baradhara, 3900 m, 23.7.1972, *Bhattacharyya* 48964 (BSD); Dadarptoo, 3400 m, 4.9.1961, *Nair* 16604 (BSD); Opp. Darcha Nala, 4000 m, 23.7.1972, *Bhattacharyya* 48942 (BSD); Darcha, 3400 m, 18.7.1979, *Aswal*10339 (CDRI); Darcha, 3500 m, 18.7.1979, *Aswal* 10365 (CDRI); Darcha (after crossing bridge), 3500 m, 4.8.2003, *Chaudhary & Khan* 225019 (LWG); Kinnaur, Sangla, 3200 m, 1.10.1964, *Nair* 34452 (BSD); On way to Reswal from Sangla, 3200 m, 7.8.1973, *Janardhanan* 52555 (BSD); Dhanachunala, 2400 m, 26.7.1963, *Malhotra & Nair* 28389 (BSD); Sangla kanda, 3100 m, 22.9.1964, *Nair* 34168 (BSD); Chitkul hill slope, above village, 22.7.1974, *Janardanan* 53642 (BSD); Baspa valley, 2743 m, September 64, *Brandis* 3811 (CAL); Right bank of Baspa, 28.7.1890, *Lace* 442 (CAL); Baspa valley, Mangsu Pass, 4115 m, 18.7.1939, *Sherifff* 7448 (BM); Rakcham, Baspa valley, 25.5.1972, *Janardhanan* 47352 (BSD); Rakcham, 3810 m, 12.8.1973, *Janardhanan* 52654 (BSD); Lahul, Baralacha La, 4400 m, 4.8.2003, *Chaudhary & Khan* 225026 (LWG); Spiti, Kunzam Pass, 4600 m, 7.9.1961, *Nair* 16691 (BSD); Lahul, *Jeschke* 270/6 (CAL); On way to Raswal from Sangla, 3200 m, 7.8.1973, *Janardhanan* 52555 (CAL); Kunawar, Pangi, 3048 m, 18.8.1847, *Thomson* (CAL); Pangi, *Stoliczka s. n.* (CAL). **Jammu & Kashmir**: Pirpas/Banihal ridge, 3000 m, 14.9.1958, *Rao* 7698 (BSD); Rajdhan, 11672 ft., 17.9.86, *Wadhwa* 83830 (BSD); Dagwan, 3500 m, 11.9.1979 *Wadhwa* 66958 (BSD); Seshnag, 3600 m, 25.7.1966, *Nair* 36962 (BSD); Lidderwal hill top, 3160 m, 1956, *Rao* 645 (BSD); Lidderwal, 3353 m, 17.7.1956, *Rao* 645 (BSD); Lidderwal 3353 m, 25. 8. 1956, *Polunin* 56/497 (BM); Synthen village, 3500 m, August 1986, *Vohra & Naithani* 83121 (BSD); Kashmir, 2134–2438 m, 27.8.1892, *Duthie* 12592 (DD); Kashmir, 3048–3353 m, 20.08.1893, *Duthie s. n.* (CAL); Kashmir, 2377 m, 20.7.1925, *Zambert* 50 (DD); Dagwan, 3500 m, 10.9.1979, *Wadhwa* 66928 (BSD); Ladakh, Dras, 3110 m, 17.7.76, Collector ? 58700 (BSD); Dras, 3048 m, 3.8.1891, *Gammie* (CAL); Ladakh, Sankoo, 3100 m, 21.8.1976, *Wadhwa & Vishwanathan s.n.* (BSD); 5 km after Sankoo towards Panikhar, 3096 m, N 34° 13.784' E 75° 58.558', 4.8.2011, *Chaudhary et al.* 263863 (LWG); Sind valley, below Baltal, 2743–3048 m, 27. 06. 92, *Duthie* 11559 (CAL); Musjid valley, 3657–3962 m, 25.9.1893, *Duthie* 13202 (BM); Kolahoi, 3353–3657 m, 8.8.1893, *Duthie* 13508 (BM); Kashmir, July 1956, *Rao* 645 (CAL); Kamri, near Gurikot, 3048–3353 m, 26.8.1892, *Duthie* 12570 (CAL); Above Liddar, 3353 m, 1.8.1893, *Duthie* 13531 (CAL); Cashem, Pashnvari, 2743 m, August 1905, *Meebold* 1377 (CAL); pahlgam, Aru, 2134–2743 m, *Thackeray* 1957 (BM); Kunshwan, 2286 m, 21.7.1876, *Clarke* 29387 (BM); Rajpariyar sanctuary, upper Bringhi, 2895 m, 20.8.1943, *Ludlow & Sheriff* 9337 (BM); Thajwas, near Sonamarg, 3200 m, 13.8.1940, *Ludlow & Sheriff* 7915 (BM); Sonamarg, 2591 m, 2.7.1978, *Lancaster* 67 (BM); Sonamarg, 2743 m, 24.8.1875, *Clarke* 27190B (BM); Nichinai, 3505 m, 2.9.1956, *Polunin* 56/649 (BM); Baltal to Matayam, Sept. 1880, *Young s. n.* (BM). **Uttarakhand**: Mandakini valley, Rambara-Garurchatti, 2700–3300 m, 14.9.1973, *Mehrotra* 4722 (CDRI); Uttarkashi, Har Ki Dun, on way to Jamdhar glacier, ± 3350 m, 28.8.1987, *Agarwal* 84352 (BSD); Uttarkashi, on way to Barasupars,

August 1996, *Balodi* 92244 (BSD); Uttarkashi, Kedarganga valley, 3600 m, 30.8.1983, *Bhattacharyya* 74636 (BSD); Uttarkashi, way to Jamnidhar glacier, 23.9.1995, *Balodi* 89597 (BSD); Garhwal, valley of flowers, 3800 m, 19.8.1963, *Bhattacharyya* 29565 (BSD); Garhwal, Mana-Ghastoli, 3500 m, 16.10.1959, *Rau* 10555 (BSD); Garhwal, Lakshmiban, 3500 m, 10.9.1959, *Rau* 10531 (CAL); Tehri, Bizula, 5.7.1999, *Pundir* 94711 (BSD); Garhwal, Lakshmiban, 3500 m, 10.10.1959, *Rau* 10531 (BSD); Garhwal, Bhyundar valley, 3600 m, 4.10.1962, *Bhattacharyya* 24341 (BSD); Tehri, Chittora, 3657 m, 29.9.1948, *Koez* 22043 (DD); Tehri-Garhwal, way to Kushi Kalyan, 2800–2900 m, 24.9.74, *Vishwanathan* 54931 (BSD); Tehri, on way to Khatling, 3500 m, *Goel* 67933 (BSD); Tehri, Khatling, 3500 m, 14.8.1978, *Goel* 64495 (BSD); Tehri-Garhwal, Binh valley, 3657–3962 m, 27.7.1883, *Duthie* 982 C (CAL); Nanda Devi, Ramni to Deodi, 27.8.1981, *Hajra* 73375 (BSD); Chamoli, North of Bampa, ± 3800 m, 6.9.1975, *Naithani* 56091 (BSD); Kumaon, Furkia/Pindari, 3500–5000 m, 23–24.9.1957, *Rao* 4577 (BSD); Kumaon, Moraine in Mulapa Gadb Darwa, 3353–3657 m, 5.8. 1886, *Duthie* 5465 (CAL); Almora, Binsar, 26.6.1996, *Basu* 76389 (BSD); Without Locality: Himal. Bor. Occ. , 2438–3962 m, *Thomson* (CAL).

Notes: The species exhibits variations in length of plants, hairs, pedicel, calyx lobes, bracts and stamens. The hairs on stems, branches, petioles, rachis and peduncles are appressed and forwardly oriented while in some specimens (e.g. *Chaudhary* 220654) they are comparatively longer and distinctly spreading. Baker (1876) has mentioned this character of hairs (spreading nature) in *A. maddenianus* Benth. ex Baker which has been now synonymised under *A. himalayanus* Klotzsch. The pedicels vary from 0.5–2 mm long even on the same plant or on different plants in same population. The bracts are quite variable in length as well as in shape. They may be smaller than pedicel or almost equal to pedicel. Usually, the calyx lobes are shorter than tube while in many specimens the lobes are almost equal to the tube. The stamens are usually monadelphous (completely closed), but sometimes diadelphous (vexillary stamen completely free from staminal sheath) condition is also seen.

5. ***Astragalus oxyodon*** Baker in Hook. f., Fl. Brit. India 2: 125. 1876; Ali, Biologia 7: 39. 1961 & in Nasir & Ali, Fl. W. Pakistan 100: 178. 1977; Sanjappa, Legum. India 92. 1992 (as *A. oxydon*); Kumar & Sane, Legum. South Asia: Checkl. 236. 2003; L.B.Chaudhary *et al.*, Taiwania 52 (3): 222, f. 7 & 11D. 2007; Xu & Podlech in Wu *et. al.*, Fl. China 10: 362. 2010; Podlech & Zarre, Tax. Rev. gen. *Astragalus* (Legum.) Old World 2: 922. 2013. –Type: West Tibet, *Falconer s. n.* (holo K!: photo LWG!).

Plants perennial herbs, prostrate, densely caespitose, stems slender, up to 7–20 cm long, with internodes longer than stipules, glabrescent to pubescent with white, appressed hairs. Stipules 2–2.5 × c. 1 mm, free from petiole, connate at base or up to half of the length on the opposite side of petiole, ovate-lanceolate with acuminate tip, glabrescent. Leaves 2.5–3 cm long, imparipinnately compound; petiole 5–7 mm long; rachis and petiole pilose with white, appressed hairs; leaflets 6–8 pairs, 3–4 × c. 1.5 mm, oblong or oblong-elliptic, cuneate at base, obtuse or retuse at apex, pilose with white, appressed hairs below, glabrescent to thinly pilose with white, spreading hairs above. Inflorescence axillary, peduncled raceme, 2–4-flowered;

peduncle 10–15 mm long, shorter than subtending leaf, pubescent with appressed, white and mixed with few black hairs. Bracts 2–3 mm long, longer than pedicel, persistent, glabrescent. Flowers 12–14 mm long, bluish-purple; pedicel c. 1.5 mm long, pubescent with light brownish to blackish hairs. Calyx c. 6 mm long, persistent, campanulate-tubular, pilose or thinly pilose with light brown, appressed hairs outside, glabrous inside, tube 2–3 mm long, oblique at mouth, lobes 3–4 mm long, usually longer than or sometimes equal to tube, linear. Standard 12–4 mm long, emarginate at apex; wing petals 9–10 mm long, diatinctly shorter than standard and keel petals; keel petals equal to or slightly exceeding the standard. Stamens monadelphous; staminal sheath c. 10 mm long, closed, apical free filaments c. 3 mm long. Ovary hairy, stipe c. 2 mm long, more or less equal to calyx tube. Pods 4–7 × 2–4 mm, elliptic-oblong, turgid, tapring at tip into a beak, with membranous valves, unilocular, pilose with white or mixed white and black spreading hairs, stipe 1.5–2 mm long, folded, more or less equal to calyx tube.

Flowering & Fruiting: July–September.

Distribution: INDIA: Himachal Pradesh, CHINA (Xinjiang)?; NEPAL; PAKISTAN.

Specimens examined: **Himachal Pradesh**: Mandi, Nargu Sanctuary, Phengadhar, 3700 m, 23.9.1996, *Singh* 1899 (LWG).

Notes: Xu & Podlech (2010) have recorde it from China, but Podlech & Zarre (2013) have not mentioned specimens from China.

18. *Astragalus* sect. *Lithophilus*

Bunge, Mém. Acad. Imp. Sci. Saint Pétersbourg 11 (16): 29. 1868; Podlech & Zarre, Tax. Rev. gen. *Astragalus* (Legum.) Old World 1: 269. 2013. –Lectotype species: *Astragalus lithophilus* Kar. & Kir.

Perennial herbs, with well developed stems, hairy with basifixed, spreading, long, white hairs, also with short black hairs in Inflorescence. Stipules free or rarely shortly adnate to petiole. Leaf rachis marcescent. Racemes short to long peduncled, few–to many-flowered. Bracts persistent. Calyx campanulate to tubular. Petals yellow; wing petals shorter or rarely longer than standard. Pods subsessile to stipitate, bilocular, glabrous to hairy.

Key to the Species

1a. Leaves 8–14 (–30) cm long; leaflets 16–18 pairs, hairy on both sides; calyx 11–13 mm long; pods 14–20 × c. 6 mm, oblong, narrowed at both ends .... **1. *A. kashmirensis***

1b. Leaves 5–6 cm long; leaflets 7–9 pairs, glabrous on upper side, hairy on lower side; calyx 15–19 mm long; pods 20–25 × 19–20 mm, ovoid-ellipsoide, inflated .... **2. *A. munroi***

1. ***Astragalus kashmirensis*** Bunge, Mém. Acad. Imp. Sci. Saint Pétersbourg 11 (16): 30. 1868 &15 (1): 34. 1869; Baker in Hook. f., Fl. Brit. India 2: 127. 1876; Ali, Biologia 7: 43. 1961 & in Nasir & Ali, Fl. W. Pakistan 100: 166. 1977; Sanjappa, Legum. India 85, 90. 1992. Kumar & Sane, Legum. South Asia: Checkl. 232. 2003; L.B.Chaudhary *et al.*, Taiwania 52 (3): 219, f. 4 & 11B. 2007; Podlech & Zarre, Tax. Rev. gen. *Astragalus* (Legum.) Old World 1: 270. 2013. –Type: India, in alpibus caschmirensibus, *Jacquemont* 778 (lecto P!; iso K!, LE *n. v.*, MSB *n. v.*).

A. kashmirensis. var. *falconeri* Baker in Hook. f., Fl. Brit. India 2: 127. 1876. –Type: India, Kashmir, *Falconer s. n.* (holo K!).

A. turgidus Roa & Balodi, Proc. Nat. Acad. Sci. India 59 (B), 4: 475. 1989. –Type: India, Jammu & Kashmir, Banihal mountains (facing upper munda), 2300 m, 14.9.1958, *Rao* 7655 (holo BSD!).

Plate 67

Plants erect or ascending, c. 35 cm long, stems elongated, branched, villous (shaggy) with fine, long, spreading, grey hairs, internodes not much elongated, crowded and completely covered with stipules in lower portion, slightly longer than stipules in middle portion and shorter than or equal to stipules in upper portion. Stipules 8–22 × 3–10 mm, free or slightly adnate to petiole at base, membranous, longitudinally multi-veined, lanceolate with acuminate tip and ciliate margins with hairs sitting on little tubercules, sparsely to moderately hairy outside with long, thin, spreading, grey hairs , glabrous inside. Leaves 8–14 (–30) cm long, imparipinnately compound; petiole 2–3 (–10) cm long, hairy as stem; rachis 5–10 cm long, grooved above, hairy as stem; leaflets 16–18 pairs, 9–15 × 5–10 mm, close, subsessile, oblong, oblong-elliptic or ovate, acute or obtuse-apiculate at apex, densely villous below and along margins, sparsely villous above. Inflorescence axillary, peduncled capitate raceme (head); head compact, many-flowered; peduncle (2–) 4.5–9 cm long, villous with long, spreading, grey or mixed with few blackish-brown hairs. Bracts 8–9 × 1.5–2 mm, distinctly longer than pedicel, persistent, membranous, prominently multi-veined, narrowly triangular or narrowly ovate with acuminate tip, densely ciliate along margins with long, spreading, grey hairs. Flowers 17–22 mm long, yellow; pedicel 3–3.5 mm long, villous with long, spreading, mixed grey and blackish- brown hairs. Calyx 11–13 mm long, tubular, tube 6–9 mm long, villous outside, glabrous inside, lobes 4.5–5 long, slightly shorter on dorsal side, linear, villous on both sides, hairs long, spreading, blackish-brown. Standard 17–22 × c. 8 mm, obovate with emarginate tip, claw not very distinct; wing petals c. 15.5 mm long, shorter than standard, lamina c. 7.5 × 3 mm, oblong with obtuse tip, upper auricle c. 1.2 × 1 mm long, oblong with obtuse tip, claw c. 8 mm long; keel petals c. 13 mm long, lamina c. 5 × 3 mm, convex on lower margins but tip not incurved, upper auricle minute, claw c. 8 mm long. Staminal sheath c.10 mm long, oblique at tip, free filaments 1–2 mm long, alternately shorter and longer; vexillary filament c. 11 mm long. Ovary c. 4 mm long, with c. 3 mm long stipe, very densely sericeous with grey hairs; style c. 6 mm long, slightly incurved towards tip, hairy on dorsal margin in lower portion, glabrous in upper portion; stigma capitate, glabrous. Pods 14–20 × c. 6 mm, oblong, turgid, narrowed at both ends, with c. 3 mm long stipe, completely bilocular, villous with spreading, grey, long hairs.

Flowering & Fruiting: June–September.

Habitat: In rocky fields. Altitudes 1800–3800 m.

Distribution: INDIA: Jammu & Kashmir - Endemic. It has been erroneousouly reported from Pakistan by Kumar & Sane (2003). Podlech & Zarre (2013) have also not reported it from Pakistan.

Specimens examined: **Jammu & Kashmir**: Kashmir, *Jacquemont s. n.*, (P); Kashmir, *Jacquemont* 778 (P); Kashmir, 1864, *Falconer* 422 (CAL); Banihal mountains (facing upper munda), 2300 m, 14.9.1958, *Rao* 7655 (BSD); Banihal mountain, facing Banihal valley, 1850–3835 m, 13.9.1958, *Rao* 7583 (BSD, CAL-student herb.); Ladakh, Kargil to Panikhar, 29.7.1988, *Chowdhery* & *Uniyal* 85902 (BSD); Tragbol, 2743 m, 16.9.1986, *Wadhwa* 83765 (BSD); Pissu Ghati, Seshnag, 3800 m, 26.6.1959, *Rao* 9712 (BSD); Apharwat, 3800 m, 21.6.1959, *Rao* 9644 (BSD, CAL-student herb); Jai valley, 2400 m, 30.5.1959, *Rao* 9082 (BSD); Near Gulmarg, 2438–2743 m, 13.6.1892, *Duthie s. n.* (CAL-student herb.); Gulmarg, 2438–2743 m, 24.6.1892, *Duthie s. n.* (CAL); Gulmarg, 3300 m, 4.8.1966, *Nair* 37105 (BSD); Limbar Nullah, Kajnag range, 2438–2743 m, 18.5.1892, *Duthie* 11036 (CAL).

2. *Astragalus munroi* Benth. ex Bunge, Mém. Acad. Imp. Sci. Saint Pétersbourg 11 (16): 30. 1868 & 15 (1): 35. 1869; Baker in Hook. f., Fl. Brit. India 2: 128. 1876; Ali, Biologia 7: 43. 1961 & in Nasir & Ali, Fl. W. Pakistan 100: 180. 1977; Sanjappa, Legum. India 91. 1992; Kumar & Sane, Legum. South Asia: Checkl. 234. 2003; Podlech & Zarre, Tax. Rev. gen. *Astragalus* (Legum.) Old World 1: 271. 2013. –Type: India, Ladakh, Pass east of Leh, 13000 ft., *Strachey s. n.* (holo K!).

Plates 22, 68

Plants prostrate, 30–70 cm long, stems branched or unbranched, very densely villous (shaggy) with white, long, spreading hairs. Stipules 5–16 × c. 1 mm, free, membranous, narrowly lanceolate, acuminate at apex, densely villous along margins and tip, otherwise glabrous. Leaves 5–6 cm long, imparipinnately compound; petiole (10–) 15–28 mm long, moderately villous, hairs similar to stem; rachis 3–3.5 cm long, grooved above, villous, hairs similar to stem; leaflets 7–9 pairs, subsessile, opposite, 15–21 × 4–6 mm, oblong, oblong-obovate or oblong-ovate, cuneate at base, entire and ciliate (villous) along margins, generally obtuse, sometimes subtruncate or retuse at apex, glabrous above, moderately villous below with white, long, spreading hairs. Inflorescence axillary, shortly peduncled raceme, 2–4-flowered; peduncle 5–10 mm long, villous. Bracts 6–10 mm long, membranous, linear, villous outside especially along margins and tip, glabrous inside. Flowers 19–21 mm long, yellow; pedicel 2–2.5 mm long, villous with white, spreading hairs. Calyx 15–19 mm long, tubular, sparsely villous outside with white spreading hairs, glabrous inside, membranous, tube 8–9 mm long, lobes 7–11 mm long, linear-setaceous, ciliate. Standard c. 21 mm long, lamina c. 13 × 9 mm, obovate, emarginated at apex; wing petals c. 20 mm long, lamina c. 9.5 × 4 mm, narrowly ovate , obtuse at apex, upper auricle c. 3 × 2.5 mm, claw c. 10.5 mm long; keel petals c. 17.5 mm long, lamina c. 7 × 5 mm, oblong, curved towards apex, obtuse at apex, upper auricle minute, claw

c. 10.5 mm long. Staminal sheath c. 16 mm long, obtuse at apex, free filaments 3–4 mm long, alternately shorter and longer. Ovary c. 11 mm long, densely hairy, with c. 3 mm long stipe; style c. 5 mm long; stigma capitate, glabrous. Pods 20–25 × 19–20 mm, yellowish-white or brownish, ovoid-ellipsoide, inflated, dorsally grooved, membranous, finely veined, with 4–5 mm long curved stipe, abruptly narrowed at apex into a c. 5 mm long beak, villous with white, spreading hairs, completely bilocular, many-seeded. Seeds c. 2.5 × 1.5 mm, reniform, subturgid, brown, with smooth surface, glabrous.

Flowering & Fruiting: July–September.

Habitat: On sandy or gravely barren soils on mountain slopes. Altitudes: 2900–5600 m.

Distribution: INDIA: Himachal Pradesh, Jammu & Kashmir; CHINA.

Specimens examined: **Himachal Pradesh:** Lahul-Spiti: Jispa, 3500 m, 15.8.1982, *Aswal* 10942 (CDRI); Lahaul, 4267 m, 29 .9.1888, *Baksh* (CAL); Before Pitsio, 3724 m, N 32° 44.564′ E 77° 14.971′, 3.8.2014, *Chaudhary et al.* 259366 (LWG); Patscio, 3962 m, 8.7.1941, *Bor* 15155 (DD); Patscio, 3962 m, 3.7.1941, *Bor* 13162 (DD); 12 km before Darcha from Patscio side, 9.8.2003, *Chaudhary* & *Khan* 225057 (LWG). **Jammu & Kashmir:** Ladakh, Mulbekh, Treaty road, 3048 m, 4. 4. 1941, *Ludlow* & *Sherriff* 8362 (BM); Ladakh, Namikala, 4000 m, 21.8.1986, *Aswal* & *Goel* 16206 (CDRI); Ladakh, Chorten Chen, 5181 m, August 1931, *Koelz* 2692d (CAL); Chorten, Chen, 5181 m, 23.8.1931, *Koelz* 2692 (DD); Lama Yuru, 4267 m, 30.8.1931, *Koelz* 2763 (DD); Mulbekh-Shergul Road, 3310–3200 m, 20.7.1976, *Wadhwa* 58803 (BSD); Budhkarbu, 3800 m, 3.9.1975, *Viswanathan* 55604 (BSD); Mathoo-Phu, 45000–4800 m, 6.8.1976, *Wadhwa* 59618 (BSD); Near Fotula, 13000 ft., Aug. 1966, *Singh s. n.* (BSD); Leh-Pulu, 3505–4560 m, 11.8.1976, *Wadhwa* 59764 (BSD); Dras-Zoji, 2900 m, 17.9.1970, *Bhattacharyya* 41233 (BSD); Saspole, 4200 m, 30.7.1976, *Wadhwa* 59325 (BSD); Ladakh, Lamayansu, 3353 m, July 1905, *Meebold* 1372 (CAL); Neighbourhood of Rambagh in Karang, 3657–4572 m, July 1865, *Stoliczka* (CAL).

19. *Astragalus* sect. *Malacothrix*

Bunge, Mém. Acad. Imp. Sci. Saint Pétersbourg 11 (16): 54. 1868; Podlech & Zarre, Tax. Rev. gen. *Astragalus* (Legum.) Old World 2: 1039. 2013. –Lectotype species: *Astragalus mollis* M. Bieb.

Perennial herbs, stemless or rarely with distinct stems, hairy with basifixed hairs. Stipules adnate to petiole or sometimes upper ones free, connate behind stem, sometimes upper ones or all free. Racemes peduncled, dense. Calyx tubular with oblique mouth, slightly inflated towards base, ruptured by pod. Standard often tongue-like narrow towards tip. Pods short, bilocular, hairy.

1. ***Astragalus pseudo-beckii*** Sirj. & Rech. f., Biol. Skr. Dan. Vid. Selsk. 9 (3): 40, f. 33 & 34. 1958; Ali, Kew Bull. 28 (2): 300. 1973 & in Nasir & Ali, Fl. W. Pakistan 100: 181, f. 27 G–J. 1977; Kumar & Sane, Legum. South Asia: Checkl. 237. 2003; L.B.Chaudhary & Z.H. Khan, Bull. Bot Surv. India 47 (1–4): 55, f. 2. 2005; Podlech & Maassoumi in Fl. Iranica 178: t. 106. 2010; Podlech & Zarre, Tax. Rev. gen. *Astragalus* (Legum.) Old World 2: 1120. 2013. –Type: Afghanistan, Sargundshar, 8.5.1949, *Neubauer* 1949/1180 (holo W *n. v.*).

Plate 69

Plants perennial herbs, c. 30 cm long, prostrate to ascending, stems striate, sparsely pilose, hairs completely appressed, black, intermixed with minute white ones. Stipules 5–7 × 2–3 mm, adnate to petiole at base, connate opposite the petiole, free portion triangular, acute to acuminate at tip, almost glabrous, only sometimes a few hairs present along margins. Leaves 5.5–6.5 cm long, imparipinnately compound; petiole 1–2 cm long; rachis 3–4 cm long, channeled above, hairy as stem; leaflets 7–9 pairs, 10–18 × 5–6 mm, opposite to alternate, elliptic, ovate, ovate-elliptic or oblong, cuneate at base, acute or obtuse at apex, glabrous to glabrescent above, pilose below with white, completely appressed hairs. Inflorescence 7–10 cm long, axillary, peduncled raceme, longer than subtending leaf, many-flowered; peduncle 4.5–6 cm long, pilose with appressed and mixed black and white hairs. Bracts c. 8 × 2 mm, decidous, not protruding the flower buds, narrowly lanceolate, pilose with spreading, mixed black and white hairs. Bracteole 1, 1.5–4 mm long, situated in the upper part of pedicel or at the base of calyx, linear, ciliate with black spreading hairs along margins. Flowers 16–18 mm long, nodded at maturity; pedicel 1–2 mm long, densely hairy with half spreading, black or mixed with few white hairs. Calyx c. 9 mm long, campanulate, pilose outside with spreading, mixed black and white hairs, glabrous inside except throat region, tube c. 5 mm long, lobes unequal, linear, dorsal ones c. 2 mm long, ventral ones c. 4 mm long. Standard 16–18 mm long, obovate, lamina c. 12 × 6 mm, claw c. 4 mm long; wing petals c. 18 mm long, almost equal to standard, lamina c. 9 × 2 mm, oblong, obtuse at tip, upper auricle c. 2.5 × 1 mm, claw c. 9 mm long; keel petals c. 15 mm long, slightly shorter than standard and wing petals, lamina c. 5 × 3 mm, upper auricle minute, claw c. 10 mm long. Staminal sheath c. 14 mm long, free filaments 2–3 mm long, alternately shorter and longer; vexillary filament c. 15 mm long. Ovary c. 6 mm long, densely hairy with white and completely appressed hairs, stipe c. 6 mm long; style c. 5 mm long, glabrous; stigma capitate, glabrous. Young pods very densely tomentose with mixed black and white hairs. Pods not seen.

Flowering & Fruiting: May–July.

Distribution: INDIA: Jammu & Kashmir; AFGHANISTAN; PAKISTAN.

Specimens examined: **Jammu & Kashmir**: Gulmarg, Tragbol, 10400 ft., 20.7.1876, *Clarke* 29302B (BM).

20. *Astragalus* sect. *Ophiocarpus*

Bunge, Mém. Acad. Imp. Sci. Saint Pétersbourg 11 (16): 9. 1868; Podlech & Zarre, Tax. Rev. gen. *Astragalus* (Legum.) Old World 1: 158. 2013. –Type species: *Astragalus ophiocarpus* Bunge.

Annual herbs, branched from base, glabrous or hairy with basifixed hairs. Stipules minute, free. Racemes sessile to subsessile, 1–5-flowered. Flowers nodded. Calyx campanulate with subulate teeth. Pods sessile, linear, terete, flexuose-curved, slightly constricted between seeds, unilocular, hairy to glabrous.

1. ***Astragalus ophiocarpus*** Bunge, Mém. Acad. Imp. Sci. Saint Pétersbourg 11 (16): 10. 1868; Baker in Hook. f., Fl. Brit. India 2: 122. 1876; Boissier, Fl. Or. 2: 224. 1872; Ali, Biologia 7:30. 1961 & in Nasir & Ali, Fl. W. Pakistan 100: 153. 1977; Sanjappa, Legum. India 92. 1992; Podlech & Zarre, Tax. Rev. gen. *Astragalus* (Legum.) Old World 1: 158. 2013. –Type: Tibet occidental, 11000 ft., *Thomson s. n.* (lecto LE *n. v.*; iso K!).

A. aitchisonii Baker in Hook.f., Fl. Brit. India 2: 121. 1876. –Type: Punjab, Salt Range, *Aitchison s. n.* (K!).

Ophiocarpus aitchisonii (Baker) Podlech, Sendtnera 2: 162. 1994; Kumar & Sane, Legum. South Asia: Checkl. 252. 2003; L.B.Chaudhary & S.K.Srivast., Taiwania 52 (1): 31. 2007.

Plate 70

Plants annual herbs, branched from the base, stems up to 25 cm long, several, branched, moderately to densely strigose with white, appressed or half appressed hairs. Stipules c. 1 mm long, free from petiole, triangular, ciliate with white hairs, otherwise glabrous. Leaves 3.5–5 cm long, imparipinnately compound; petiole 10–12 mm, hairy as stem; leaflets c. 9 pairs, c. 5 × 2 mm, opposite or alternate, oblong with retuse apex, velutinous with white, spreading hairs on both sides; petiolule c. 0.5 mm long. Inflorescence axillary, almost sessile; raceme 3–5-flowered, lax. Bracts c. 1 mm long, ciliate. Flowers c. 5 mm long, subsessile, pale violet; pedicel minute, strigose with white hairs. Calyx c. 2.5 mm long, campanulate, strigose with white, appressed hairs, tube c. 1.5 mm long, lobes c. 1 mm long, linear-lanceolate. Standard 5–6 mm long, obtuse or slightly retuse at apex; wing petals 4–5.5 mm long; keel petals 4.5–5 mm long. Staminal sheath nearly truncate at apex. Ovary sessile, hairy. Pods c. 30 x 2 mm, sessile, nodded, narrowly linear, with a hooked c. 3 mm long beak, terete, flexuose or semicircularly to almost circularly curved, slightly constricted between seeds, unilocular, 10–20-seeded, appressed hairy to glabrous.

Flowering & Fruiting: March–July.

Distribution : INDIA: Jammu & Kashmir, Punjab; AFGHANISTAN; IRAN; IRAQ; PAKISTAN; TADZHIKISTAN; TURKMENISTAN.

Specimens examined: **Jammu & Kashmir:** Karakorum, Ashkoley, 3080 m, 11.8. 1876, *Clarke* 30404 (K). **Punjab**: Muktesar, 17.3.1966, Nair 36527 (BSD).

Notes: Podlech (1994) placed this species under a separate genus *Ophiocarpus* (Bunge) Ikonn. Lock and Schrire (2005) have also accepted this treatment. However, in molecular analysis it nested within *Astragalus* L. (Kazempour Osaloo *et al.*, 2003), hence again Podlech & Zarre (2013) placed it in *Astragalus*.

21. *Astragalus* sect. *Oroboidei*

A. Gray, Proc. Amer. Acad. 6: 203. 1864; Podlech & Zarre, Tax. Rev. gen. *Astragalus* (Legum.) Old World 2: 824. 2013. –Type species: *Astragalus oroboides* Hornem.

A. sect. *Hemiphaca* Bunge, Mém. Acad. Imp. Sci. Saint Pétersbourg 11 (16): 20. 1868. –Type species: *Astragalus hemiphaca* Kar. & Kir.

Perennial herbs to under shrubs, erect, stems well developed, many, hairy with basifixed hairs. Stipules free from petiole, sometimes may be connate behind the stem. Leaves imparipinnately compound with marcescent petiole. Racemes long peduncled, globose to shortly cylindric, often elongating in fruiting. Bracteoles absent. Calyx campanulate. Standard often without claw; wing petals longer than keel petals, obtuse or emarginated to biloded at apex. Pods sessile to subsessile, ovoid, grooved dorsally, unilocular or fully to partially bilocular, hairy or glabrous.

1. ***Astragalus macropterus*** DC., Prod. 2: 283. 1825; Baker in Hook. f., Fl. Brit. India : 128. 1876; Gontsch. & Boriss. in Komarov, Fl. U. R. S. S. 12: 82. 1946; Ali, Biologia 7: 32. 1961 & in Nasir & Ali, Fl. W. Pakistan 100: 158. 1977; Sanjappa, Legum. India 90. 1992; Kumar & Sane, Legum. South Asia: Checkl. 233. 2003; Podlech in Fl. Iranica 178: t. 17. 2010; Podlech & Zarre, Tax. Rev. gen. *Astragalus* (Legum.) Old World 2: 842. 2013. –Type: ex Dahuria misit Fischer, leg. *Pausner* s. n. (with young fruits) (lecto G-DC *n. v.*; iso LE *n. v.*).

A. vicioides Ledeb., Fl. Altaica 3: 301. 1831. –Type: Kazakhstan, In montibus Tarbagatai, 20.6., *von Schrenk s. n.* (holo LE *n. v.*; iso P *n. v.*)

Plants perennial herbs to under shrubs, erect, 30–100 cm long, with many stems arising from the top of thick rootstock, stems stiff, smooth or finely striate, terete below, ± angled above, thinly pubescent with very minute, appressed, white hairs, sometimes with a few black hairs only on nodal region. Stipules 1–3.5 mm long, free from petiole, spreading, caducous, narrowly triangular, sparsely pubescent outside, glabrous inside. Leaves 2–5 cm long, imparipinnately compound; petiole 1–6 mm long, thinly pubescent; rachis 1.3–4.5 cm long, filiform, channeled above, pubescent with minute appressed, white or sometimes mixed with some black hairs; leaflets 3–6 pairs, 4–10 × 0.5–1 mm, subsessile, opposite or sometimes alternate lower ones, sparse, thick, linear to linear-oblong or linear-lanceolate, upwardly curved along margins, acute at apex, glabrous above, pubescent with minute, appressed, white hairs below. Inflorescence 8–25 cm long, axillary, long peduncled raceme;

raceme many flowered, lax; peduncle 5–14 cm long, glabrescent to thinly pubescent with appressed mixed white and black hairs; rachis 8–11 cm long, pubescent with appressedly as well as spreading, minute, mixed black and white hairs. Bracts minute to 2 mm long, persistent, shorter than or ± equal to pedicels, lanceolate, hairy particularly along margins with black hairs. Flowers 9–10.5 mm long, pink or lilac mauve; pedicel 1.5–2.5 mm long, pubescent with appressed black hairs. Calyx 2–2.5 mm long, campanulate, oblique at mouth, persistent, pubescent with appressed, minute mixed black & white hairs outside, glabrous inside, lobes minute (less than 1 mm long), deltoid with pointed tip. Standard 9–10.5 × c. 4.5 mm, narrowly elliptic, emarginate at apex, without distinct claw; wing petals 8 – 9 mm long, shorter or nearly equaling the standard, lamina 5.5–6.5 × c. 1.5 mm, oblong with obtuse tip, upper auricle minute, spurs present, claw c. 2.5 mm long; keel petals 5–5.5 mm long, convex on lower margin, much shorter than standard and wing petals, upper auricle minute, spurs present, claw c. 2.5 mm long. Staminal sheath c. 3 mm long, emarginate at tip, free filaments 1–2 mm long, alternately shorter and longer; vexillary filament c. 4 mm long. Ovary c. 3 mm long, subsessile, glabrous; style c. 2 mm long, curved towards apex, glabrous. Pods 7–9 × c. 3 mm, subsessile, ovoid, acuminate at tip, turgid, reticulately veined, glabrous, completely bilocular, 5–6-seeded.

Flowering & Fruiting: June–August.

Habitat: On stony ground.

Distribution: INDIA: Jammu & Kashmir; AFGHANISTAN; CHINA; N. EURASIA; PAKISTAN.

Specimens examined: **Jammu & Kashmir:** Ladakh, Bothkhabru, 3300 m, 26.8.1986, *Aswal & Goel* 16411 (CDRI); Kargil, August 1982, *Hajra* 74053 (BSD); Kargil, 2743 m, 8.6.1977, *Stainton* 7814 (BM); Below Kangi, 2438–2743 m, July 1907, *Meebold* 1397 (CAL).

22. *Astragalus* sect. *Poliothrix*

Bunge, Mém. Acad. Imp. Sci. Saint Pétersbourg 11 (16): 48. 1868; Podlech & Zarre, Tax. Rev. gen. *Astragalus* (Legum.) Old World 2: 999. 2013. –Type species: *Astragalus leucocephalus* Grah. ex Benth.

Perennial herbs, with well developed numerous slender stems, white and black hairy. Stipules free from petiole, all or only lower ones connate opposite the petiole (behind the stem). Racemes with well developed slender peduncle. Bracts whitish-membranous. Calyx campanulate. Petals whitish to violet; wing petals longer than keel petals. Pods sessile to stipitate, straight to curved, unilocular or rarely incompletely bilocular, 2–4-seeded.

Key to the Species

1a. Leaflets hairy on both sides 2

1b. Leaflets glabrous (or sometime sparsely hairy as in *A. strictus*) above, hairy below 4

2a. Calyx 7–7.5 mm long, teeth 4–4.5 mm long; pods sessile, included into calyx .... **2. *A. leucocephalus***

2b. Calyx 3.5–5 mm long, teeth very minute; pods with 2–2.5 mm long stipe, excluded from calyx....................3

3a. Hairs subbasifixed.................... **3. *A. minutidentatus***

3b. Hairs basifixed ... . **4. *A. sanjappae***

4a. Peduncle 8.5–9 cm long; pods slightrly to discinctly curved .... **6. *A strictus***

4b. Peduncle 0.3–5 cm long; pods stright....................5

5a. Stems glabrous to sparsely hairy; stipules and bracts glabrous .... **7. *A. tenuicaulis***

5b. Stems hairy; stipules and bracts hairy 6

6a. Leaves 1–3.5 cm long; leaflets 1–4 mm long, elliptic..... **1. *A. cobresiiphilus***

6b. Leaves 3.5–6 cm long; leaflets 6 –7 mm long, narrowly obovate or oblong .... **5. *A. sikkimensis***

1. *Astragalus cobresiiphilus* Podlech & L.-R. Xu, Sendtnera 7: 196. 2001; Xu & Podlech in Wu *et al.*, Fl. China 10: 383. 2010; Podlech & Zarre, Tax. Rev. gen. *Astragalus* (Legum.) Old World 2: 1005. 2013. –Type: China, Xizang, valley c. 8 km SSE of Lhasa, 3950 m, *Dickoré* 3384 (holo MSB *n. v.*).

Plants up to 18 cm long, prostrate to ascending, caespitose, with appressed to subappressed hairs, stems numerous, with somewhat flattened white hairs, partly glabrescent to glabrous. Stipules 2–3.5 mm long, greenish or membranous, free from petiole, ovate-triangular, distinctly connate opposite the petiole, sparsely hairy to glabrous. Leaves 1–3.5 cm long; petiole short; rachis sparsely to loosely hairy like stem; leaflets 5–8 pairs, 1–4 × 0.6–2 mm, elliptic, rounded to emarginate at apex, glabrous above, sparsely to loosely covered with appressed hairs below. Inflorescence axillary, peduncled raceme; raceme 3–10-flowered; peduncle 1.5–5 cm long, sparsely to loosely hairy like stem but hairs increasingly black toward tip. Bracts 1–1.5 mm long, membranous, ovate to narrowly triangular, white and black hairy. Flowers blue to purple or rarely white. Calyx (2.5–) 3–4 mm long, campanulate, slightly oblique at mouth, loosely covered with appressed black hairs and more rarely also with some white hairs; teeth unequal, 0.3–1.5 mm long, lower three narrowly triangular, other two triangular. Standard 6–7 × c. 5 mm, widely elliptic to orbicular, deeply emarginate; wing petals c. 5.5 mm long, narrowly elliptic, obtuse at apex, claw 2 mm long; keel c. 5 mm long, obliquely elliptic, subacute at apex, curved, claw c. 2 mm long. Ovary with c. 1 mm long stipe, white hairy; style glabrous. Pods c. 6 × 2 mm, subsessile or with a c. 1 mm long stipe, narrowly ellipsoid, curved, acute with hooked persistent style, grooved dorsally, appressed black hairy, glabrescent with age, unilocular. Seeds 3–4, 2 × 1.2 mm, blackish.

Flowering & Fruiting: June.

Distribution: INDIA: Jammu & Kashmir; CHINA; NEPAL.

Specimens examined: Nil.

Notes: Recorded here after Podlech & Zarre (2013) who have reported it from India based on the specimen 'Kashmir, Ladakh, Tsakzhun Tso, 4500 m, 20.6.1931, *Koelz* 2380 (W)'.

2. *Astragalus leucocephalus* Grah. ex Benth. in Royle, Illustr. Bot. Himal. Mount. 198, t. 32, f. 2. 1835; Baker in Hook. f., Fl. Brit. India 2: 128. 1876; Ali, Biologia 7: 25. 1961 & in Nasir & Ali, Fl. W. Pakistan 100: 147, f. 19 B–G. 1977; Sanjappa, Legum. India 90. 1992; Kumar & Sane, Legum. South Asia: Checkl. 233. 2003; L.B. Chaudhary, Rheedea 15 (1): 77. 2005; Podlech & Zarre, Tax. Rev. gen. *Astragalus* (Legum.) Old World 2: 1009. 2013. –Type: India, Dehra Doon & Riskeekes, 1825, *Wallich* 5923A (lecto K!; iso CGE *n. v.*, E, PRC *n. v.*).

Plates 23, 71

Plants 10–20 cm long, prostrate to ascending, occasionally erect, stems densely caespitose, many, variable in length (up to 20 cm long), generally unbranched, slender, densely pilose with white, unequal, almost appressed hairs. Stipules 7–12 × 2–2.5 mm, longer than internodes, persistent, free from petiole, connate up to the middle opposite the petiole, triangular-ovate, acuminate at apex, densely white pilose outside, glabrous inside. Leaves 5–11 cm long, imparipinnately compound, closely arranged on stem; petiole 1.5–2 cm long, densely pilose; rachis 3.5–8 cm long, grooved above, densely pilose with white, appressed as well as spreading hairs; leaflets 21–35, 5–9 × 1.8–5 mm, usually very close, opposite, subopposite or alternate, subsessile, oblong, elliptic, ovate, narrowly ovate or obovate, obtuse, subacute, acute or sometimes retuse with a minute stalked gland at apex, cuneate at base, entire along margins, densely pilose on both surfaces or sometimes sparsely pilose above, densely pilose below, hairs white, spreading on upper surface, appressed on lower surface. Inflorescence axillary, peduncled compact capitate raceme, many-flowered; raceme 7–20 × 10–15 mm, oblong, oblong-ovate or almost globose; peduncle 1.8–12.5 cm long, terete, densely pilose with white, intermixed appressed and spreading hairs. Bracts 4–5 × 1.5–2 mm, lanceolate with acuminate tip, densely white pilose. Bracteoles 2, attached at the base of calyx, c. 1 mm long, linear, densely white pilose. Flowers pinkish-white or yellowish, 7–8 mm long, subsessile. Calyx 7–7.5 mm long, campanulate-tubular, densely pilose to shaggy outside, glabrous inside, tube c. 3 mm long, lobes 4–4.5 mm long, subequal, linear-triangular. Standard 7–8 × c. 3 mm, slightly exerted from calyx, narrowly obovate with broad basal portion, emarginate at apex; wing petals c. 6.5 mm long, lamina c. 3.5 × 1 mm, oblong- obovate with minutely dentate apex, claw c. 3 mm long, upper auricle and spurs present; keel petals 4.5–5 mm long, lamina 2–2.5 × c. 1.5 mm, curved towards apex, upper auricle minute, claw c. 2.5 mm long. Staminal sheath c. 4 mm long, truncate at apex, free filaments c. 1.5 mm long; vexillary filament c. 4 mm long. Ovary c. 3.5 × 2 mm, sessile, elliptic-ovate, densely white sericeous; style c. 2 mm long, curved from base, glabrous; stigma capitate, glabrous. Pods 4–5 × c. 2

mm, sessile, included into calyx, ellipsoid or oblong-ovoid, turgid, densely pilose with subsppressed hairs, unilocular, c. 6-seeded. Seeds 1.5–1.7 × c. 1 mm, brown, reniform, turgid, smooth, glabrous.

Flowering & Fruiting: (May–) July–September (–October).

Habitat: commonly growing on mountain slopes in grassy habitats in open areas in sandy as well as in stony/hard soils, at the altitudes ranging from 1200 m to 5500 m.

Distribution: INDIA: Himachal Pradesh, Jammu & Kashmir, Uttarakhand, Sikkim; AFGHANISTAN; NEPAL; PAKISTAN.

Specimens examined: **Himachal Pradesh**: From Panjian to Aut resthouse, 21.5.1959, *Srivastava & Party* 58562, 58584, 59425, 59416 (LWG); Simla, 2438 m, 12.9.1954, *Singh* 15805 (LWG); Simla, Kreda Das, 2591 m, 20.5.1954, *Singh* 7300 (LWG); Simla, Limestone hills, 1829 m, 19.6.1877, *Gamble* 43976 (CAL); Chamba, Bharmour, 23.7.1963, *Malhotra & Nair* 28326 (BSD); Chamba, Bharmour, 2150 m, 23.7.63, *Malhotra & Nair* 28312 (BSD); Sarahan, 2400 m, 26.5.1962, *Nair* 21903 (BSD); Mahasu dist., Rohru Paber valley, 1600 m, 30.7.1965, *Nair* 36215 (BSD); Kulu, Manikaran, 3 Kms. from Manikaran towards Phulga, 1600 m, 17.9.2001, *Chaudhary* 220770 (LWG); After 2 km from Manikaran, 1874 m, N32° 01.255′ E 77° 22.003′, 19.7.2008, *Chaudhary et al.* 249613 (LWG); Chamba, Gothan ridge, 2134 m, 27.4.1898, *Lace* 1693 (CAL); Cherat, August 1897, *Collett s. n.* (CAL); Without locality, 983 m, 27.8.1977, *Murti & Prasad* 62126 (BSD); Without locality, 2134 m, May 1845, *Thomson* (CAL). **Jammu & Kashmir**: Kishtwar, Hargad enclosure, 28.5.1986, *Uniyal* 80270 (BSD); Panchori *Cedrus* forest area, 8.6.1986, *Kumar* 81457 (BSD); Shankaracharya hill top, 2300 m, 25.7.1956, *Rao* 792 (BSD); Sopore, 1700 m, 12.6.1959, *Rao* 9425 (CAL, BSD); Sankaracharya, 1833 m, 19.5.1954, *Prasad* 10424 (LWG); Tangmarg to Baramula, near Pattan, 30.6.1957, *Kaul & Party* 23539 (LWG); Srinagar, Dathankot Road, 12th mile, 9.8.1956, *Saran & Party* 30263 (LWG); Andanwar, 13.5.1962, *Khan* 96309 (LWG); Srinagar, 1829 m, 25.7.1958, *Thapliyal* 26293 (DD); Ladakh, Mulbekh (N to NW hills), 3310 m, *Wadhwa* 58873 (BSD); Ladakh, Sanmodangsa, 3900-4100 m, 29.8.1976, *Wadhwa* 59520 (BSD); Ladakh, Pilong Chho, 3310 m, 20.7.1976, *Wadhwa* 58832 (BSD), Ladakh, Chushul, 5500 m, 24. 8. 1975, *Viswanathan* 55012 (BSD); Kashmir, 1864, *Falconer* 421 (CAL); Kashmir, July 1956, *Rao* 792 (CAL). **Sikkim:** Llonak, 4511 m, 5.8.1909, *Smith & Cave* 2142 (LBG); Chanra Gumpa, on dry hill sides, 4267 m, 5.10.1923, *Rhomoo* 7639 (LBG). **Uttarakhand**: Garhwal Himalaya: Chamoli, Suraithola, *Hajra* 87381 (BSD); Chamoli, way to Dunagiri, 2600 m, 16.8.1974, *Naithani* 54045 (BSD); Pauri Gharhwal, Chopra Pauri, 22.10.1975, *Naithani* 526 (DD); 18 km from Mussoorie on Chamba Road, 2380 m, N 30° 26.746′ E 78° 11.731′, 6.9.2006, *Chaudhary et al.* 229567 (LWG); Mussoorie, Dhobighat, 1500 m, 9.8.1967, *Bhattacharyya* 37278 (BSD); Mussoorie, Company garden, 3.4.1962, *Malhotra* 20176 (BSD); Mussoorie, Camp Bagh, 2000 m, 26.10.1967, *Arora* 37817 (BSD); Mussoorie, Laudar, 2500 m, 24.12.1956, *Sarin* 1387 (BSD); Mussoorie, 2000 m, 23.01.1977, *Aswal.* 6266 (CDRI); Mussoorie, Jhari Pani, 28.4.1932, *Ram* (DD); between Mussoorie and Kempty fall, near Himalaya Adventurous Centre, 1829 m, 15.8.2001, *Chaudhary* 220627 (LWG); Mussoorie, on way to Dehra Dun, 1400 m, 3.4.1966, *Malhotra* 36840

(BSD); Garhwal, 1864, *Falconer* 421 (CAL); Mussorie,. *King s. n.* (CAL); Jaunsar, 2591 m, April 1892, *Gamble* 23474 (CAL); Tehri-Garhwal, 1524–1830 m, April 1891, *Duthie s. n.* (CAL). Dehra Dun, Tons valley, 1200 m, 2.10.1995, *Rana* 108252 (LWG); Dehra Dun, Sahasradhara, 4.1.1964, *Malhotra* 30786 (BSD); Chakrata, Kanasar, 2743 m, 14.8.2001, *Chaudhary* 220625 (LWG); Deoban, above spider colony, 2438 m, 13.8.2001, *Chaudhary* 220620 (LWG); Jaunsar, 2134 m, May 1891, *Gamble* 22834 (DD); Kumaon Himalaya: Nainital/Ayalpata, 21000–2500 m, 8.10.1957, *Rao* 4902 (BSD); Nainital, 2286 m, 27.5.1967, *Pradhan*. (BSD); Nainital, 2134–2286 m, 2.4.1925, *Osmaston* 1245 (DD); China, 2286 m, 19.7.1913, *Gill* 591 (CAL, LWG); Ranikhet, 27.3.1968, *Balapure* 72740, 72750 (LWG); Haldwani, forest nursery, 8.5.1963, *Kaul & Party* 94645 (LWG); Nainital, 16.5.1857, collector? *s. n.* (CAL); Nainital, 2072 m, 1854, collector? s. *n.* (CAL); Nainital, Tiffin Top, 2286 m, 27.5.1967, *Pradhan* 7 (CAL); Sambal, Cashem, 1524 m, June 1905, *Meebold* 1396 (CAL); N. W. Himalaya : Naldera, 1980 m, 23.4.1886, *Johnson* (CAL); Serahan, *Stoliczka s. n.* (CAL); Dawar, 6/64, *Brandis* 3647 (CAL); Without locality, June 1898, *Gamble* 2692 (CAL).

Notes: It exhibits variations in the length of plants, size of leaflets and density of hairs. It was noticed that the variation in size is usually due to ecological factors. The plants grow in sandy and loose soils are longer than those grow in rocky and hard soils. The root is thick, hard and deep penetrating which may go up to 30 cm long, depending upon the age of plants.

3. ***Astragalus minutidentatus*** Y.C.Ho, Bull. Bot. Lab. North-East. Forest. Inst. 1(3): 108, f. 12. 1981; Xu & Podlech in Wu *et. al.*, Fl. China 10: 385. 2010; Podlech & Zarre, Tax. Rev. gen. *Astragalus* (Legum.) Old World 2: 1011. 2013. –Type: China, Xizang, Yadung, 2850 m, 11.6.1975, *Exped. ad Qinghai-Xizang* 750285 (holo PE *n. v.*; iso KUN *n. v.*).

Plants up to 30 (–50) cm long, prostrate to ascending, covered with ± flattened and subbasifixed (mostly with an acute appendix at base) hairs, stems many, up to 25 cm long, slender, densely hairy, hairs subappressed to ascending, in upper part to spreading, white, at nodes also black. Stipules 2–3 mm long, narrowly triangular to triangular, free from petiole, lower ones distinctly connate opposite the petiole, upper ones free from each other, with white and some black hairs. Leaves 2.5–5 cm long, subsessile; petiole c. 5 mm long, white hairy, rachis subappressed white hairy; leaflets 6–10-paired, 3–7 × 1.5–3.5 mm, elliptic, rounded to emarginate at apex, loosely covered with subappressed, flexuous, tangled hairs on upper surface densely covered with subappressed, ± straight, white hairs on lower surface. Inflorescence axillary, raceme, densely 6–8-flowered; peduncle 1–2 cm long, with ascending to spreading white and black hairs. Bracts 1.5–2 mm long, white and black hairy. Flowers lilac, pale mauve or white; pedicel c. 1 mm long, black hairy. Calyx c. 4 mm long, tubular-campanulate, slightly obliquely gibbous at base, slightly oblique at mouth, subappressed black hairy; teeth minute, unequal, triangular. Standard 8–10 × 3.5–4 mm, obovate, emarginated at pex, claw indistinct; wing petals 8–10 mm long, lamina 5–6 × 1.8–2 mm, narrowly elliptic, obtuse at apex, auricle less than 1 mm long, claw 4-5 mm long; keel petals 7–9 mm long, lamina 4–5 × 2–2.5 mm, obliquely elliptic, obtuse at apex, with widely curved lower edge and nearly straight to slightly concave upper edge, auricle minute, claw 4–5 mm long. Staminal tube truncate at apex. Ovary with c. 2 mm long stipe, white hairy. Pods 5–7 × c. 3 mm, ± erect, obliquely ellipsoid, grooved dorsally, with c. 2 mm long stipe and c. 1 mm long beak, unilocular, 2–3-seeded, loosely to rather densely covered with subappressed to slightly ascending black hairs, sometimes also with white hairs when young. Seeds c. 2 × 1.5 mm, brown, dull.

Flowering & Fruiting: August.

Habitat: Altitudes 4270–4580 m.

Distribution: INDIA: Uttarakhand: BHUTAN, CHINA; NEPAL.

Specimens examined: **Uttarakhand:** Tehri Garhwal, Nila valley, 4270–4580 m, 17.8.1883, *Duthie* 988 (CAL).

Notes: Podlech & Zarre (2013) have reported it from India based on specimen 'Tehiri Garhwal, Nila Valley, 4270–4580 m, 17.8.1883, *Duthie* 988 (BM)'. Same specimen was also observed at CAL.

4. ***Astragalus sanjappae*** L.B.Chaudhary & Z.H.Khan, Rheedea 15 (2): 129, f. 1. 2005; Podlech & Zarre, Tax. Rev. gen. *Astragalus* (Legum.) Old World 2: 1018. 2013. –Type: India, Sikkim, North Sikkim District, Thangu, Yongdi, after military camp, 4200 m, 13.8.2004, *Chaudhary* 225071 (holo LWG!; iso CAL!).

A. jumlaensis Podlech, Feddes Repert. 117: 227. 2006. –Type: Nepal, Bhurchula Lehh, near Jumla, 82°10′ E, 29°17′ N, 3820 m, 11.7.1952, *Polunin, Sykes* & *Williams* 4526 (holo BM *n. v.*).

Plate 72

Plants prostrate, (5–) 15–36 cm long, with fibrous or woody rootstock, several stems arising from the base, stems slender with distant internodes (internodes 2.5–5 cm in the middle portion of stem), pubescent with flat, white, appressed to half spreading hairs, sometimes black hairs present at nodal region. Stipules 2–3 (– 5) × 1.0 –1.5 mm, free from petiole, connate at base on the opposite side of the petiole or free, spreading, triangular-lanceolate, pilose with white hairs outside, glabrous inside. Leaves 2.5–5.5 cm long, imparipinnately compound; petiole 0–10 mm long; rachis and petiole pilose with appressed, generally white or sometimes mixed with few black hairs; leaflets 5–9 pairs, 3–9 × 1.5–4 mm, alternate (in lower portion) to opposite (in upper portion), ovate, narrowly ovate, oblong or oblong-elliptic, acute, obtuse, emarginate or retuse at apex, strigose with white, appressed hairs below, thinly pilose with spreading, white hairs above. Inflorescence axillary, capitate raceme, shortly peduncled; peduncle 0.3–1.5 (–2.5–3.5) cm long, shorter than subtending leaf, pilose with black or white or mixed black and white, spreading or appressed hairs; raceme 4–12 (–16)-flowered, dense. Bracts 2–3 (–4) × c. 1 mm, almost equal to or exceeding the pedicel, lanceolate, pilose with long, mixed black and white or black hairs. Flowers 8–11 mm long, purple or white with purple tinge; pedicel 1–2 mm long, generally densely pubescent with black hairs. Calyx 3.5–5 mm long, campanulate-tubular, very oblique at mouth, pilose with black or sometimes mixed black and white, appressed, flat hairs outside, glabrous inside, more hairy on teeth region, teeth very minute (sometimes indistinct). Standard 8–11 × c. 4 mm, narrowly elliptic, emarginate at apex; wing petals c. 10 mm long, usually shorter than or sometimes almost equal to standard, lamina c. 5 × 1.5 mm, oblong, obtuse, claw c. 5 mm long; keel petals 9–9.5 mm long, shorter than wing petals, lamina c. 4.5 × 2 mm, oblong with slightly incurved towards rounded tip, claw c. 5 mm long. Staminal sheath 7–8 mm long, obtuse at apex, free filaments 1.5–2 mm long; vexillary filament 8–8.5 mm long. Ovary 2–2.5 mm long, stipitate (stipe c. 2.5 mm long), c. 6-ovuled, pubescent with white, appressed hairs; style c. 4 mm long, incurved at tip, glabrous; stigma capitate, glabrous. Pods (5–) 6–7.5 × 3–4 mm, stipitate, oblong-ellipsoid, narrowed at apex, turgid, grooved dorsally, unilocular, (2–) 4 (–6)-seeded, pubescent with minute, flat, spreading, generally black or sometimes white or mixed black and white hairs, stipe 2–2.5 mm long, shorter than calyx tube. Seeds c. 2 × 1.5 mm, reniform, turgid, smooth, glabrous, black.

Flowering & Fruiting: (April–) July–September.

Habitat: In sandy or hard soils on moist mountain slopes, road sides, river sides and flat lands at altitudes ranging from 3300–4200 m.

Distribution: INDIA: Sikkim, Uttarakhand; NEPAL.

Specimens examined: **Sikkim**: North Sikkim District: Thangu, 4000 m, 12.8.2004, Chaudhary 225066 (LWG). **Uttarakhand**: Garhwal Himalaya, Uttarkashi District: 4 km away from Gangotri towards Chirbasa, 3360 m, 5. 9. 2002, *Khan* 223081 (LWG); Chirbasa, 3500 m, 5 September 2002, *Khan* 223085 (LWG); Bhojwasa, 3700 m, 5.9.2002, *Khan* 223100 (LWG); Bhojwasa, 3700 m, 11. 9.2002, *Khan* 206792 (LWG); Gangotri-Bhojwasa, 31.7.1987, *Uniyal & Balodi* 80536 (BSD); Chirbasa to Bhojwasa, 3343 m, N 30° 58.827′ E 79° 01.855′, 4.9.2006, *Chaudhary et al.* 229546 (LWG); Gangotri to Chirbasa, 3206 m, N 30° 59.658′ E 78° 59.375′, 4.9.2006, *Chaudhary et al.* 229559 (LWG); Pushpabasha-Gaumukh, 3900 m, 24.9.1967, *Naithani* 37437 (BSD); Rhudughera, 3048–3353 m, 19.7.1883, *Duthie* 986b (BM, CAL).

5. ***Astragalus sikkimensis*** Benth. ex Bunge, Mém. Acad. Imp. Sci. Saint Pétersbourg 11 (16): 23. 1868 & 15 (1): 28. 1869; Baker in Hook. f., Fl. Brit. India 2: 126. 1876; Grierson & Long in Grierson & Long, Fl. Bhutan 1 (3): 721. 1987; Sanjappa, Legum. India 94. 1992; Press *et. al.*, Anno. Checkl. Fl. Pl. Nepal 167. 2000; Kumar & Sane, Legum. South Asia: Checkl. 240. 2003; Podlech & Zarre, Tax. Rev. gen. *Astragalus* (Legum.) Old World 2: 1019. 2013. –Type: India, Sikkim, Lachen, Samdong, 8000–12000 ft., *Hooker s. n.* (lecto P!; iso C *n. v.*, CAL!, CGE *n. v.*, E *n. v.*, G *n. v.*, GOET *n. v.*, K!, L *n. v.*, LD *n. v.*, M *n. v.*, MSB *n. v.*, STU *n. v.*, W *n. v.*).

Plates 24, 73

Plants caespitose, prostrate, 10–70 cm long, stems several, reddish, elongated, slender, with distant internodes, glabrous in lower portion, moderately pubescent with minute, white, appressed, flat hairs in upper portion. Stipules 2.5–3 × 0.7–1.5 mm, free from petiole, connate only at base opposite the petiole, spreading, ovate-lanceolate. Leaves 3.5–6 cm long, imparipinnately compound; petiole 3–5 mm long; rachis pubescent with white, flat, appressed hairs; leaflets 6–9 pairs, 6–7 × 2–4 mm,opposite or alternate below and opposite above, narrowly obovate or oblong, emarginated at apex, glabrous above, pilose with silky, appressed hairs below. Inflorescence axillary, capitate or loose raceme, 4–14-flowered; peduncle 0.5–3 cm long, reddish, shorter than subtending leaf, thinly pubescent with minute, appressed, blackish-brown hairs. Bracts c. 2 mm long, linear or narrowly ovate with acute tip, white or black hairy. Flowers 9–10 mm long, white with light purple tinge or purple; pedicel c. 2 mm long, pubescentwith black hairs. Calyx 4–5 (–6) mm long, campanulate, reddish, tube 2.5–3 (–4) mm long, hairy outside with blackish-brown, appressed hairs, lobes 1.5–2.5 mm long, shorter than or sometimes equal to tube, hairy on both sides. Standard 9–10 mm long, narrowly elliptic-obovate, emarginate at apex; wing petals 8–9 mm long, shorter than standard, lamina 4–5 × c. 1 mm, oblong, obtuse at apex, claw c. 4 mm long; keel petals almost equal to wing petals, lamina 4–5 × c. 2 mm, onlong, curved in upper part, obtuse at apex, claw c. 4 mm long. Staminal sheath c. 6 mm long, truncate at apex, free filaments 1.5–2 mm long; vexillary filament c. 7 mm long. Ovary with c. 2 mm long stipe, densely hairy with light grey, appressed hairs. Pods 9–11 × 4–5 mm, partially reddish, oblong-ellipsoid, stipitate, turgid, grooved dorsally, densely pubescent with minute, blackish-brown or black, spreading hairs, unilocular, usually c. 5 seeded, stipe c. 3 mm long, equal to calyx tube. Seeds c. 1.5 × 1 mm, reniform, turgid, yellowish or blackish-brown, smooth.

Flowering & Fruiting: July–August.

Habitat: Common (particularly in Lachung region of Sikkim) on grassy hill slopes, along roadsides, flat lands, barren lands or river beds in rocky or sandy soils. Altitudes 2500–5000 m.

Distribution: INDIA: Himachal Pradesh, Jammu & Kashmir, Sikkim; BHUTAN; CHINA; NEPAL.

Specimens examined: **Himachal Pradesh:** Dharamshala, near Ilaqa, 3200 m, 16.9.2017, *Bhatia s. n.* (LWG). **Sikkim**: North Sikkim: Lachung proper, Near bridge, 2800 m, 16.8.2004, *Chaudhary* 225087 (LWG); Lachung, after 4 km towards Yungthang, 3100 m, 15.8.2004, *Chaudhary* 225078 (LWG); Lachung, 2 km away from Yungthang, near Garampani, 3800 m, 15.8.2004, *Chaudhary* 225083 (LWG); Between Lachung and Yungthang, inside Singma *Rhododendron* sanctuary, 3500 m, 15.8.2004, *Chaudhary* 225082 (LWG); Lachung valley, Yungthang, 3962 m, 22.8.1892, *Gamme s. n.* (CAL); Yungthang, 3962 m, 22.08.1892, *Gammie* 635 (CAL); Lachung-Yungthang, 3700 m, 26.7.1990, *Kumar & Singh* 9589 (BSHC); Lachung to Dombeyang, 2500 m, 13.7.1987, *Raju & Singh* 7569 (BSHC); On way to Lachung, 26.7.1989, *Srivastava* 12368 (SIKKIM); Near Kanchanjungha, August 1887, *King's collector* (CAL); Lachung, near Kanchanjungha, August 1887, *King's collector* (CAL); Lachung valley, Onglathang, 4572 m, July 1888, collector ? *s. n.* (CAL); Near Bansoi 17.8.1989, *Srivastava* 10227 (BSHC); Pa to Yongdi, 22.7.1988, *Raju* 8846 (BSHC); Tankra Int., 3353 m, 05.8.1892, *Gammie* 635 (CAL); Nijan, 1888, *King's collector* (CAL); Zemu valley, 3596 m, 11.7.1909, *Smith & Cave* 1209 (CAL, LBG); Llonak, 3962 m, 30.7.1909, *Smith & Cave* 2775 (CAL); Singbil, 3962–4267 m, 11.8.1900, *Smith* 4169 (CAL, LBG); Changu, 3048 m, 22.9.1926, collector ? 176 (CAL); Thangu to Gochung, 15.7.1986, *Raju* 5989 (BSHC); Thangu to Gochung, 11.7.1986, *Raju* 5879 (BSHC); Thangu-Muguthang, 20.7.1997, *Das* 19314 (BSHC); Lachen-Thangu, 17.8.1989, *Mondal* 10521 (BSHC); Lachen, 2743 m, 22.10.1923, *Rohmoo Lepcha* 2014 (LBG); Between Lachen and Thangu, 2906 m, N 27° 48.603′ E 88° 32.388′, 4. 9. 2015, *Chaudhary* & *Kushwaha* 259376 (LWG); Lachen, Zeema, 3387 m, N 27° 49.229′ E 88° 32.951′, 4. 9. 2015, *Chaudhary* & *Kushwaha* 259381 (LWG); Before Thangu, Kelep, 3800 m, 12.8.2004, *Chaudhary* 225063 (LWG); Thangu, 4000 m, 12.8.2004, *Chaudhary* 225069 (LWG); Towards Nathula, 2. 9. 2015, *Chaudhary* & *Kushwaha* 259387 (LWG).

Notes: I could not observe the specimens from Jammu & Kashmir. Podlech & Zarre (2013) have recorded it from Jammu and Kashmir based on following specimen: Kashmir, Khelanmarg, 3300 m, 11. 8. 1956, *Polunin* 56/174, 56/182 (BM).

6. *Astragalus strictus* Grah. (in Wall., Cat. 5924. 1831–32, *nom. nud.*) ex Benth. in Royle, Illustr. Bot. Himal. Mount. 198. 1835; Baker in Hook. f., Fl. Brit. India 2: 124. 1876; Ali, Biologia 7: 41. 1961 & in Nasir & Ali, Fl. W. Pakistan 100: 160. 1977; Grierson & Long in Grierson & Long, Fl. Bhutan 1 (3): 720. 1987; Press *et al.*, Annot. Checkl. Fl. Pl. Nepal 167. 2000; Kumar & Sane, Legum. South Asia: Checkl. 241. 2003; Podlech & Zarre, Tax. Rev. gen. *Astragalus* (Legum.) Old World 2: 1020. 2013. –Type: Mountain of Sillet, *Wallich Cat.* 5924 (holo K–W *n. v.*).

Astragalus ladakensis Balakr., J. Bomaby Nat. Hist. Soc. 63: 328. 1966; Sanjappa, Legum. India 94. 1992. –Type: that of *A. strictus* Grah. ex Benth., non Siev. ex Fisch. 1825.

Plates 25, 74

Plants procumbent or slightly decumbent, deep rooted, densely caespitose, reaching up to 26 cm long, stems several, slender, branched, with long internodes, sparsely to moderately pubescent with minute, white, appressed hairs. Stipules 3–5 × 1–2 mm, free from petiole, connate up to the middle or sometimes only at base opposite the petiole, deltoid or ovate- lanceolate with subacuminate to acuminate tip, sparsely pilose with white appressed hairs outside, glabrous inside, ciliate along margins. Leaves 4–9 cm long, imparipinnately compound; petiole 7–18 mm long, longer in basal leaves; rachis 3.5–8 cm long, channeled above, pubescent with white appressed hairs; leaflets 6–10 pairs, 4–8 × 2–4 mm, subsessile, alternate, subopposite or opposite, obovate, ovate, elliptic or oblong, cuneate at base, entire along margins, subobtuse to obtuse or obtuse-retuse at apex, glabrous to unevenly sparsely pilose above, pilose with white hairs below (hairs more along midrib region). Inflorescence (6–) 9–11 cm long, axillary, peduncled condensed raceme; raceme 1.5–3 × 1.5 cm, globose or oblong, many-flowered, compact; peduncle (4–) 8.5–9.5 cm long, striate (in dry condition), pubescent with white or mixed white and black appressed hairs, floral rachis pubescent with completely black or mixed black and white hairs. Bracts 1.5–2.5 mm long, longer than pedicel, persistent, deflexed in lower flowers. Flowers 7–10 mm long, purple, pale purple, blue or white, deflexed or half deflexed at maturity; pedicel less than 1 mm long, densely pubescent with black or mixed black and white hairs. Calyx 3–5 mm long, campanulate, persistent, pilose outside with black or mixed black and white, appressed hairs, glabrous inside, tube 1.5–3 mm long, lobes 1.5–2 mm long, equal to or slightly shorter than tube. Standard 7–10 mm long, narrow at base, obovate with emarginate tip, wing petals 6–9 mm long, shorter than standard, lamina 4–6 × c. 2 mm, almost oblong with obtuse tip, upper auricle and spurs present, claw 2.5–3 mm long; keel petals 5–7 mm long, shorter than wing petals, lamina c. 4 × 2.5 mm, obovate, acute at tip, in upper part curved on ventral edge, spurs and upper auricle present, claw c. 3 mm long. Staminal sheath c. 5 mm long, obtuse at tip, free filaments 2–3 mm long; vexillary filament c. 7 mm long. Ovary c. 2 mm long, with c. 2 mm long stipe, pubescent with mixed black and white, appressed hairs; style 2.5–3 mm long, curved, glabrous; stigma capitate, glabrous. Pods 3–6 × 1–3 mm, shortly stipitate, narrowly oblong, slightly to distinctly curved, gooved dorsally, narrowed at tip into a minute beak, unilocular, pubescent with black or white or mixed black and white appressed hairs, 4–8-seeded.

Flowering & Fruiting: (June–) July–September.

Habitat: On rocky slopes, dry open plains, along agriculture fields and irrigation canal and in open meadows. Altitudes 3500–5500 m.

Distribution: INDIA: Himachal Pradesh, Jammu & Kashmir, Sikkim, Uttarakhand; BHUTAN; CHINA; NEPAL; PAKISTAN.

Specimens examined: **Himachal Pradesh**: Lahul-Spiti, 12 km before Sarchu, 4479 m, N 32° 50.451′E 77° 28.966′, 30.7.2011, *Chaudhary et al.* 263818 (LWG); Spiti, Kiato, 4100 m, 29.7.1972, *Bhattacharya* 49110 (CAL, BSD); Spiti, Losar, 4100 m, 26.7.1972, *Bhattacharyya* 48886 (CAL, BSD); Lossar, Near PWD guest house, 4073 m, N 32° 26.322′ E77° 45.374′, 6.8.2012, *Chaudhary & Bajpai* 259341 (LWG); Spiti, La Darcha Plateau, 4300, 30.7.1972, *Bhattacharyya* 49151 (CAL, BSD); Spiti, Chandratal, 4169 m, 8.9.1961, *Nair* 16762 (BSD). **Jammu & Kashmir**: Ladakh: Yeru after Upsi, 3550 m, 6.8.2003, *Chaudhary & Khan* 225043 (LWG); Leh, Shey, 3380 m, 7.8.2003, *Chaudhary & Khan* 225049 (LWG); Rumtse, 4112 m, N 33° 38.024′ E 77° 44.766′, 31.7.2011, *Chaudhary et al.* 263830 (LWG); Leh, Phey, 3400 m, 7.8.2003, *Chaudhary & Khan* 225050 (LWG); Ladakh, 1.8.1976, *Wadhwa* 59411 (BSD); Sultak, 5500 m, 1.9.1975, *Vishwanathan* 55091 (BSD); Shyok, 4400 m, 30.8.1975, *Vishwanathan* 55061(BSD); Mathoo-Phu, 4500–4800 m, 6.8.1976, *Wadhwa* 59621 (BSD); Chang La, 17350 ft., 27.8.76, *Wadhwa* 60094 (BSD); Tangste, Lukung Road, 3875 m, 28.8.76, *Wadhwa* 60138 (BSD); Tangtse, 4400 m, 28.8.75, *Vishwanathan* 54777 (BSD); Ganglas, 7.8.1988, *Chowdhery & Uniyal* 86044 (BSD); Ganglas Gompha, 3749 m, N 34° 12.356′ E 77° 36.745′, 1.8.2011, *Chaudhary et al.* 263833 LWG). **Sikkim**: North Sikkim: Giagong to Koreng, 4877–5181 m, 1. 8. 1987, *Raju & Singh* 7992, 7995 (BSHC); Gaigong plains, 30.7.1987, *Raju & Singh* 7958 (BSHC); Giagong plains, 30.7.1987, *Raju & Singh* 7951 (BSHC); Giagaon, 4600 m, 13.8.2004, *Chaudhary* 225074 (LWG); Giagaon, 4359 m, N27° 56.691′ E88° 33.889′, 4.9.2015, *Chaudhary* & Kushwaha 259384, 259385 (LWG); Muguthang, towards Tebla Pass, 31.7.1988, *Rai* 9400 (BSHC). Llonak, 3962 m, 26. 7. 1909, *Smith & Cave* 1774 (LBG); Sepo La, 5181 m, 14. 8. 1909, *Smith & Cave* 2396 (CAL); Gyangtse, July–September 1904, *Walton* 108 (LBG); Cholamoo, 4572 m, August 1885, *King's collector Gyalu* (CAL); Wthout locality, 4572 m, *Hooker* (CAL). **Uttarakhand**: Kumaon, Nikchang valley, 3962–4267 m, 31.8.1884, *Duthie* 2808 (CAL, K).

7. ***Astragalus tenuicaulis*** Benth. ex Bunge, Mém. Acad. Imp. Sci. Saint Pétersbourg 11 (16): 23. 1868; & 12: 28. 1869; Baker in Hook.f., Fl. Brit. India 2: 126. 1876; Ali, Biologia 7: 41. 1961; Sanjappa, Legum. India 95. 1992; L.B.Chaudhary *et al.*, Taiwania 52 (3): 223, f. 9 & 11F. 2007; Podlech & Zarre, Tax. Rev. gen. *Astragalus* (Legum.) Old World 2: 1024. 2013. –Type: India, prov. Sikkim pr. Lachen, Samdong, 7000–9000 ft., *Hooker* (lecto P *n. v.*; iso BM *n. v.*, G *n. v.*, K!, L *n. v.*, W *n. v.*, P *n. v.*).

A. inconspicuus Baker in Hook.f., Fl. Brit. India 2: 126. 1876. –Type: India, Kumaon, banks of Ralam river, 9000 ft., *Strachey & Winterbottom s. n.* (holo K!).

Plates 26, 75

Plants prostrate, caespitose, up to 40 cm long, stems partially reddish, very slender, with distant internodes, almost glabrous to sparsely pubescent with white, minute, appressed hairs. Stipules c. 1.5 mm long, glabrous, other details similar to *A. sikkimensis.* Leaves 3.5–40 cm long, imparipinnately compound; petiole c. 7 mm long; rachis thinly pubescent with minute, white, appressed hairs; leaflets 5–7 pairs, oblong, narrowly obovate or oblong-elliptic, obtuse or retuse at apex, glabrous above, thinly hairy with minute, flat, white, appressed hairs below. Inflorescence axillay, 2–9-flowered raceme or capitate raceme; peduncle 2.5–13 mm long, shorter than subtending leaf, glabrescent to thinly pubescent with minute, appressed, white hairs. Bracts c. 2 mm long, longer than or more or less equal to pedicel, glabrous. Flowers 7–9 mm long, purple; pedicel 1–2 mm long, glabrescent to thinly pubescent with minute, appressed, white hairs. Calyx 4–5 mm long, campanulate-tubular, glabrescent to sparsely pubescent with minute black, appressed hairs, oblique at mouth, tube c. 3 mm long, lobes 1–1.5 mm long. Standard 7–11 mm long, longer than wing and keel petals; wing petals 6–10 mm long, equal to keel petals. Pods 7–9 × 3–4 mm, oblong, partially tinged with purple colour, sessile to subsessile (stipe less than or c. 1 mm long), turgid, unilocular, thinly or sometimes moderately pubescent with minute, completely appressed, generally black or sometimes grey or brownish hairs. Seeds 1.5 × 1 mm, other details similar to *A. sikkimensis.*

Flowering & Fruiting: July–August (–October).

Habitat: Very occasional on open hill slopes in sandy moist soil along road sides.

Distribution: INDIA: Sikkim, Uttarakhand. Endemic.

Specimens examined: **Sikkim:** North Sikkim: Choongtong, 1524 m, 28.7.1892, *Gammie* 299 (CAL); Zema to Log-bridge, 2.10.1997, *Singh & Das* 19906 (SIKKIM); Lachen, towards Thangu near Zema, 3000 m, 12. 8. 2004, *Chaudhary* 225060 (LWG). **Uttarakhand**: Kumaon, banks of Ralam river, 3000 m, *Strachey & Winterbottom s. n.* (K).

Notes: Podlech & Zarre (2013) have recorded it from Bhutan without citing any specimen.

23. *Astragalus* sect. *Pseudosesbanella*

Podlech, Feddes Repert. 117. 230. 2006; Podlech & Zarre, Tax. Rev. gen. *Astragalus* (Legum.) Old World 1: 252. 2013. –Type species: *Astragalus hoffmeisteri* (Klotzsch) Ali.

Perennial herbs, with well developed erect or prostrate stems. Hairs basifixed. Stipules free, green. Racemes long peduncled, lax. Bracts persistent. Bracteoles minute. Calyx campanulate. Petals more or less of same length. Stigma hairy. Pods stipitate, compressed laterally, unilocular to bilocular.

Key to the Species

1a. Plants erect and almost glabrous; pods septum less than 1 mm broad... **2. *A. hoffmeisteri***

1b. Plants prostrate and pubescent with spreading hairs; pods septum more than 1 mm broad .... **1. *A. falconeri***

1. ***Astragalus falconeri*** Bunge, Mém. Acad. Imp. Sci. Saint Pétersbourg 11 (16): 4. 1868 & 15 (1): 2. 1869; Baker in Hook.f., Fl. Brit. India 2: 120. 1876; Ali, Biologia 7: 14. 1961; Bot. Notiser 118: 87. 1965 & in Nasir & Ali, Fl. W. Pakistan 100: 130. 1977; Sanjappa, Legum. India 87. 1992; Kumar & Sane, Legum. South Asia: Checkl. 228. 2003 p. p. (excluding *A. hoffmeisteri* var. *pilosus* Ali); Podlech & Zarre, Tax. Rev. gen. *Astragalus* (Legum.) Old World 1: 252. 2013 p. p. (excluding *A. hoffmeisteri* var. *pilosus* Ali).

Plants perennial herbs, prostrate, spreading, tufted, stems terete, finely striate, tomentose with white, spreading hairs in younger portions, glabrescent in older portion. Stipules 3–5 × 2–3 mm, foliaceous, free, thick, broadly ovate, acuminate at tip, ciliate with white spreading hairs along margins, glabrescent to sparsely pilose on both sides. Leaves 2–6 cm long, imparipinnately compound; petiole minute to 5 mm long; rachis c. 4.5 cm long, pilose with white spreading hairs; leaflets 8–9 pairs, c. 4 × 2 mm, subsessile, subopposite to alternate, obovate, cuneate at base, entire along margins, retuse, truncate or obtuse at apex, hairy on both sides or glabrous on upper side, hairy on lower side. Inflorescence (2.5–) 6–13 cm long, longer than subtending leaf, axillary, peduncled raceme, few to many-flowered, laxed; peduncle 5.5–7 cm long, pilose with white, spreading hairs; floral rachis 1–6 cm long, pubescent with white, appressed hairs. Bracts 1.5–2 mm long, almost equal to or slightly shorter than pedicel, membranous, persistent, linear-lanceolate, pilose. Bractoles minute, situated at the base of calyx, membranous, ciliate. Flowers 11–13 mm long, bluish-pink; pedicel 1.5–2.5 mm long, pubescent with white, spreading hairs. Calyx c. 5 mm long, campanulate, persistent, tube c. 3 mm long, oblique at mouth, glabrous to glabrescent outside, glabrous inside, lobes subequal, sparsely pilose both sides, ciliate along margins, linear-lanceolate, upper ones c. 1.5 mm long, lower ones c. 2 mm long. Standard 11–13 mm long, backwardly deflexed, lamina 9–11 × c. 8 mm, broadly elliptic, retuse at apex, claw c. 2 mm long; wing petals c. 9.5 mm long, slightly shorter than standard & keel petals, lamina c. 6.5 × 2.5 mm, oblong with subobtuse tip, upper auricle c. 1 mm long, claw c. 3 mm long; keel petals 11–13 mm long, ± equal to standard, lamina 8–10 × 4.5 mm, abruptly incurved in upper part, upper auricle c. 1 mm long, claw c. 3 mm long. Staminal sheath c. 7 mm long, free filaments 2.5–3.5 mm long, alternately shorter and longer; vexillary filament c. 8 mm long. Ovary c. 4 mm long, linear, hairy throughout or thinly hairy only along upper suture, with c. 3 mm long stipe; style c. 4.5 mm long, abruptly incurved; stigma penicillate. Pods 14 – 16 × c. 4 mm, stipitate, oblong-ellipsoid, turgid, straight or slightlhy curved, pointed at both ends, dorsally grooved, faintly veined, glabrous or pilose with white spreading hairs, partially bilocular, stipe 5–7 mm long, longer than calyx. Seeds c. 3 × 2 mm, reniform, brown, smooth, glabrous.

1a. var. *falconeri*. –Type: West Tibet, *Falconer s. n.* (holo K!).

Plates 27, 76

Leaflets sparsely to moderately pilose with white, spreading hairs below, glabrescent to sparsely pilose above with same type of hairs. Ovary c. 4 mm long, linear, hairy. Pods pilose with white, spreading hairs.

Flowering & Fruiting: July–September.

Habitat: In the exposed moist sandy-rocky soils in the agriculture fields and on mountain slopes among herbs. Altitudes 3000–4000m.

Distribution: INDIA:Jammu & Kashmir; AFGHANISTAN; CHINA; PAKISTAN; TADZHIKISTAN.

Specimens examined: **Jammu & Kashmir**: Ladakh, Kargil, 2800–3000 m, 5.9.1976, *Wadhwa & Vishwanathan* 60198 (BSD); Suru valley, Sankoo, 3100–4100 m, 18.8.1976, *Wadhwa* 59807, 59804 (BSD); Suru valley, Panikhar, 3300–4000 m, 24.8.1976, *Wadhwa* 59875 (BSD); Kargil to Panikhar, 29.7.1988, *Chowdhery & Uniyal* 85905 (BSD); Near Panikhar village, 3263 m, N 34° 06.722′ E 75° 55.853′, 4.8.2011, *Chaudhary et al.* 263868 (LWG); Gudhai valley, 3353–3657 m, 22.7.1892, *Duthie s. n.* (CAL); Astor, 2438–2743, 22.8.1892, *Duthie* 12502 (CAL).

1b. var. *pilosus* (Ali) L.B.Chaudhary, Taiwania 52 (3): 218, f. 2 & 11A. 2007.

A. hoffmeisteri var. *pilosus* Ali, Kew Bull. 1958 (2): 304. 1958; Biologia 7: 15. 1961& in Nasir & Ali, Fl. W. Pakistan 100: 132. 1977. –Type: India, Karakorum, 10500 ft., 6.8.1876, *Clarke* 30111A (holo K!: photo LWG!).

Leaflets glabrous above, hairy below. Ovary and young pods thinly hairy only along upper suture. Pods gabrous.

Flowering & Fruiting: June–August.

Distribution: INDIA: Jammu & Kashmir. Endemic.

Specimens examined: **Jammu & Kashmir**: Baltistan, Hushe Nallah, Khane, 2895 m, 18. 6. 1828, *Ludlow* 361 p. p. - specimen marked as B (BM).

Notes: This variety distinctly differs from typical variety *falconeri* in above mentioned charctres. However, Wenninger (1991) and Podlech & Zarre (2013) have reduced it as synonym under *A. falconeri var. falconeri*.

2. ***Astragalus hoffmeisteri*** (Klotzsch) Ali, Kew Bull. 1958: 304.1958; Biologia 7: 15. 1961 & in Nasir & Ali, Fl. W. Pakistan 100: 131. 1977; Wenninger, Mitt. Bot. Staatssamml. Munchen 30: 70, f. 6 c-d, 20 a-d, 27 c. 1991; Sanjappa, Legum. India 89. 1992; Podlech in Rech. f., Fl. Iranica 174: t. 47. 1999; Kumar & Sane, Legum. South Asia: Checkl. 230. 2003; Podlech & Zarre, Tax. Rev. gen. *Astragalus* (Legum.) Old World 1: 254. 2013.

Phaca hoffmeisteri Klotzsch in Klotzsch & Garcke, Bot. Ergebn. Pr. Waldemar 159. 1862. –Type: Bot. Erg. Pr. Waldemar, tab. 2. 1862 (neo).

A. adesmifolius Benth. ex Bunge, Mém. Acad. Imp. Sci. Saint Pétersbourg 11 (16): 4. 1868 & 15 (1): 2. 1869; Baker in Hook. f., Fl. Brit. India 2: 121. 1876. –Type: that of *Phaca hoffmeisteri.*

Plates 28, 77

Plants perennial erect herbs, c. 60 cm. high, branched, stems & branches terete, glabrous. Stipules 3–8 × 2.5–5 mm, free, persistent, encirching the stem, thick, obliquely cordate-ovate, entire along margins, acute at apex, glabrous except tip portion where some minute and white hairs found along margins. Leaves 4.5–8.5 cm long, imparipinnately compound; petiole 3–5 mm long; rachis 4–8 cm long, channeled above, sparsely pubescent with minute, white, appressed hairs, glabrescent at maturity; leaflets 15–18 pairs, 2–6 × 1.5–5 mm, reducing in size towards the upper portion, thick, obovate to broadly obovate, cuneate at base, entire along margins, emarginate, retuse or truncate at apex, glabrous on both sides or sometimes minutely pubescent along apex and midrib region on lower surface; petiolules less than 1 mm long, glabrous. Inflorescence 5–26 cm long, axillary, peduncled raceme, many-flowered, very lax; peduncle 2–7.5 cm long; floral rachis 3–18.5 cm long, glabrous to glabrescent, hairs minute, white, appressed. Bracts 1.5–2 mm long, linear-oblong with acute tip, with glabrous surfaces and minutely hairy margins. Flowers 9–10 mm long, nodded, white, yellow, yellowish-green, yellowish-purple, light purple, pink, blue or greenish- blue; pedicel c. 1.5 mm long, pubescent with minute, white, appressed hairs. Calyx c. 4 mm long, persistent, campanulate, glabrous or with a few hairs on margins of lobes, tube c. 2.5 mm long, lobes unequal, lower ones 1.5 mm long, upper ones c. 1 mm long, all triangular-lanceolate. Standard 8–9 mm long, obovate; wing petals 8–9 mm long, lamina oblong-obovate with acute to subobtuse tip, claw c. 3 mm long; keel petals 9–10 mm long, abrupty incurved towards tip. Ovary c. 4 mm long, glabrous, with c. 4 mm long stipe; style c. 3 mm long, abruptly incurved from base; stigma penicillate. Pods 10–17 × 4–7, oblong or oblong-elliptic, flat, narrowed at both ends, membranous, glabrous, faintly veined, dehiscent, 2–7-seeded, partially bilocular, septum poorly developed, less than c. 1 mm broad, stipe c. 6 mm long, distinctly longer than calyx. Seeds c. 2.3 × 2 mm, reniform, flat, smooth, glabrous.

Flowering & Fruiting: June–August.

Habitat: *Generally* on hill slopes in sandy and rocky soils in open places. Altitudes 2500–4500 m.

Distribution: INDIA: Jammu & Kashmir; AFGHANISTAN; CHINA; PAKISTAN.

Specimens examined: **Jammu & Kashmir**: Ladakh, Kargil to Panikhar, 29.7.1988, *Chowdhery & Uniyal* 85904 (BSD); Near Kargil town, 3042 m, N 34° 31.272′ E 76° 07.905′, 4.8.2011, *Chaudhary et al.* 263873 (LWG); Near Kargil, August 1966, *Singh s. n.* (BSD); Kargil, 3200 m, 15.9.1970, *Bhattacharyya* 41169 (BSD); 5 km after Kargil town towards Dras, 2705 m, N 34° 36.105′ E 76° 03.918′, 5.8.2011, *Chaudhary et al.* 263874 (LWG); Ladakh, way to Fotula, 3600 m, 21.8.1986, *Aswal & Goel* 16223 (CDRI); Khalsi-Saspole Road, 2700 m, 26.7.1976, *Wadhwa* 59211 (BSD); Khalsi-Saspole Road, ± 2700 m, 26.7.1976, *Wadhwa* 59227 (BSD); Tangtse, 4500 m, 20.8.1975, *Vishwanathan* 54741 (BSD); 31.8.1975 *Vishwanathan* 55076 (BSD); on way to Khardungla, 3900 m, 9.8.1980, *Bhattacharyya* 71545 (BSD); Ladakh, Kharbu, 2500 m, 4.8.1980, *Bhattacharyya* 71394 (BSD); Nimu, 4.8.1995, *Balodi* 91080 (BSD); Ladakh, Shribichan, 3048 m, Aug., 1905, *Meebold* 1393 (CAL-Student herb.); Nubra valley, Diskit, 3200 m, 16.8.1998, *Goraya et al.* 1095 (BSD); Baltistan, Shingo valley, 3048–3353 m, 8.7.1892, *Duthie s. n.* (BM); without precise locality and date, *Wadhwa*, 59425 (BSD).

Notes: In this species the stipules are free while Ali (1977) has mentioned united stipules.

24. *Astragalus* sect. *Pseudotapinodes*

Podlech & L. R. Xu, Sendtnera 7: 199. 2001; Podlech & Zarre, Tax. Rev. gen. *Astragalus* (Legum.) Old World 2: 1034. 2013. –Type species: *Astragalus orbicularifolius* P.C.Li & C.C.Ni.

Perennial herbs, stemless to shortly stemed. Stipules adnate to petiole or connate in front of petiole. Leaflets small. Calyx campanulate. Pods incompletely bilocular.

1. ***Astragalus rigidulus*** Benth. ex Bunge, Mém. Acad. Imp. Sci. Saint Pétersbourg 11 (16): 23. 1868 & 12: 25. 1869; Baker in Hook.f., Fl. Brit. India 2: 123. 1876; Grierson & Long in Grierson & Long, Fl. Bhutan 1 (3): 720. 1987; Sanjappa, Legum. India 94. 1992; Press *et. al.*, Anno. Checkl. Fl. Pl. Nepal 167. 2000; Kumar & Sane, Legum. South Asia: Checkl. 240. 2003; Podlech & Zarre, Tax. Rev. gen. *Astragalus* (Legum.) Old World 2: 1039. 2013. –Type: India, Sikkim, in alpibus Sikkimensibus [Tungu], 13000 ft., *Hooker s. n.* (lecto P *n. v.*; iso G–DC *n. v.*, K *n. v.*, L *n. v.*, P *n. v.*, W *n. v.*, CAL!: photo LWG!).

Plants perennial herbs, 7–10 cm high above the ground, rhizomatous, rhizome slender, branched below the surface, stem up to 5 cm long, glabrous. Stipules 3–5 mm long, membranous, ovate, subobtuse at apex, adnate to the petiole up to the middle, shortly connate opposite the petiole, sometimes also connate infront of petiole, glabrous or ciliate along margins. Leaves c. 6 cm long, imparipinnately compound; petiole c. 3 cm long, glabrous to glbrescent; leaflets 8–10 pairs, c. 7 × 1 mm, linear-blong or narrowly ovate, obtuse or subacute at apex, glabrous on upper side, thinly coated with white, appressed hair on lower side. Inflorescence axillary, peduncled capitate raceme, 5–10-flowered; peduncle c. 10 mm long, distinctly shorter than subtending leaf. Bracts c. 3 mm long. Flowers 10–12 mm long, purple-red; pedicel c. 3 mm long. Calyx 5–6 mm long, campanulate, tube 2.5–3 mm long, pubescent with black, appressed hairs outside, glabrous inside, teeth equal to tube, lanceolate, black hairy inside. Standard 10–12 mm long, spatulate, emarginated at apex, longer than wing petals; wing petals 7–9 mm long, longer than keel petals, narrowly oblong with rounded to slightly emarginate apex; keel c. 6 mm long. Ovary subsessile, densely black hairy. Pods 10–15 × c. 6 mm, sessile, oblong-ellipsoid, shortly acuminate, turgid, sparsely appressed black hairy, glabrous at full maturity, unilocular, 6–8-seeded.

Flowering & Fruiting: May–June.

Habitat: Altitudes 3800–4400 m.

Distribution: INDIA: Sikkim; BHUTAN; CHINA.

Specimens examined: Examined only by type specimen.

Notes: It is a very rare species. After the collection of Type specimens no other collections have been made of this species. The species was also not noticed during my recent visit to North Sikkim. It is close *to A. alpinus* L. and chiefly characterized by very short peduncle and sessile pods.

25. *Astragalus* sect. *Rhacophorus*

Bunge, Mém. Acad. Imp. Sci. Saint Pétersbourg 11 (16): 77 & 83. 1868; Podlech & Zarre, Tax. Rev. gen. *Astragalus* (Legum.) Old World 2: 1413. 2013. –Lectotype species: *Astragalus compactus* Lam.

A. sect. *Stenonychium* Bunge, Mem. Acad. Sci. Saint Petersbourgser. 11 (16): 77 & 79. 1868. –Lectotype species: *Astragalus arnacantha* M.Bieb.

A. sect. *Microthrix* Sirj., Repert. Spec., Nov. Rregni Veg. 47: 197. 1939. –Lectotype species: *Astragalus microcephalus* Willd.

A. sect. *Macrothrix* Sirj. in Repert. Spec. Nov. Rregni Veg. 47: 225. 1939. –Lectotype species: *Astragalus barbajovis* DC.

Perennial small shrubs, densely branched, stems in older portion covered by old rachises and stipules. Hairs basifixed. Stipules adnate to petiole, shortly connate or free from each other. Leaves paripinnately compound, dense, rachis thick, rigid, straight, spiny at tip, spreading. Inflorescence axillary, cluster of 2–30 flowers, dense. Calyx turbinate, membranous, splitting with age along main veins, densely hairy. Standard constricted in middle; wing petals more or less equal to standard; keel petals shorter than wing petals; claws of wings and keels adnate to staminal sheath. Pods ovoid to ellipspoid, dorsiventrally compressed, with hooked apical beak, densely hairy, unilocular, one-seeded.

1. *Astragalus verus* Olivier, Voy. emp. Othoman. 3: tab. 44. 1807; Kumar & Sane, Legum. South Asia: Checkl. 244. 2003; Podlech & Zarre, Tax. Rev. gen. *Astragalus* (Legum.) Old World 2: 1528. 2013. –Type: Iran, inter Kermanschah *et* Hamadan Persiae occidentalis mediae, *Olivier* & *Bruguière s. n.* (holo P *n. v.*).

A. strobiliferus Benth. in Royle, Illustr. Bot. Himal. Mount. 199. 1835; Baker in Hook. f., Fl. Brit. India 2:135. 1876; Parker, For. Fl. Punjab 140. 1924; Ali, Biologia 7: 18. 1961; Univ. Studies (Karachi) 3 (3): 40. 1966 & in Nasir & Ali, Fl. W. Pakistan 100: 194. 1977; Sanjappa, Legum. India 95. 1992. –Type: India, Kanum in Kunawur, *Royle s. n.* (holo LIV *n. v.*; iso LE: India, herb. *Royle n. v.*).

Tragacantha strobilifera (Benth.) Kuntz, Revis. Gen. 2: 948.1891.

Astracantha strobiliferus (Benth.) Podlech., Mitt. Bot. Staatssamml. Munchen 19: 20.1983.

Plates 29, 78

Plants perennial woody low shrubs, erect, generally cushion forming, with thick woody root stock, profusely branched, stems hard, densely armed with ascending, spine-tiped, stramineous, persistent rachis, internodes minute and not visible as completely covered with persistent stipules, densely wooly with white hairs in young stage. Stipules 6–8 × c. 2 mm, yellowish, persistent, adnate to petiole up to the middle, membranous, ovate, acuminate at tip, faintly veined, tomentose with white hairs in young stage, glabrous at maturity. Leaves 30–50 mm long, paripinnately compound; petiole c. 10 mm long, broaden at base; rachis almost terete, smooth, hard, spiny at tip, petioles and rachis tomentose with white hairs in young stage and glabrous at maturity; leaflets 5–8 pairs, 5–8 × c. 1 mm, thick, caducous, oblanceolate, narrowly obovate or elliptic-oblong, cuneate at base, entire along margins, acute or subobtuse with spiny-micro at apex, tomentose with white, spreading hairs on both sides in young stage, glabrescent at maturity. Flowers c. 10 mm long, yellowish-white, sessile, axillary, scarcely exceeding the stipules. Bracts c. 5.5 × 1 mm, boat shaped, densely wooly with white hairs. Calyx c. 6 mm long, persistent, 5-lobed, cleft nearly to the base, tube c. 1 mm long, membranous, glabrous inside, lobes c. 5 mm long, equal, linear, very densely wooly with white hairs and collectively looke like a bunch of hairs. Standard c. 10 × 3.5 mm, panduriform (distinct in young flowers), lamina c. 7 × 3.5 mm, oblong, constricted in upper portion, emarginated at apex, claw c. 3 × 1.5 mm (not very clearly demarcated from lamina in matured flowers); wing petals c. 10 mm long, equal to standard, lamina c. 3.5 × 1 mm, oblong with obtuse tip, upper auricle minute, spurs present, claw c. 7 mm long; keel petals c. 9 mm long, equal to slightly shorter than standard and wing petals, lamina c. 3 × 1.5 mm, upper auricle minute, spurs present, claw c. 6 mm long. Staminal sheath c. 7 mm long, free filaments 2–2.5 mm long; vexillary filament c. 8 mm long. Ovary c. 4 mm long, linear, sessile, densely sericeous; style c. 5 mm long, filiform, slightly curved in upper portion, glabrous; stigma capitate, glabrous. Pods (after Podlech & Zarre, 2013) 4–6 × 2–3 mm, ovoid to ellipsoid, with a straight c. 1.5 mm long beak, densely tomentose towards base, villous upward.

Flowering & Fruiting: July–September.

Habitat: Common in open on dry, hard and rocky flat grounds and hill slopes. Altitudes 2700–4300 m.

Distribution: INDIA: Himachal Pradesh, Jammu & Kashmir AFGHANISTAN; IRAN; PAKISTAN; TURKMENISTAN; TADZHIKISTAN.

Specimens examined: **Himachal Pradesh**: Spiti, Ki-Kaza, 3800 m, 4.8.1972, *Bhattacharyya* 49318 (BSD, CAL-student herb.); Kaza, 16.9.2001, *Srivastava* 98988 (BSD); Rangrik Bridge, 3800 m, 5.8.1972, *Bhattacharya* 49334 (BSD, CAL); Rangrik, 3700 m, 4.8.2002, *Chaudhary & Khan* 206773 (LWG); Shego, 3500 m, 5.8.2002, *Chaudhary & Khan* 206776 (LWG); 2 km before Sumling (12 km before Kaza from Lossar), N 32° 16.106′ E78° 00.821′, 3725 m, 7.8.2012, *Chaudhary & Bajpai* 259350 (LWG); Near Lari town, N32° 04.976′ E78° 25.174′, 3363 m, 8.8.2012, *Chaudhary & Bajpai* 259351 (LWG); Kanam (Bashahr), 4000 m, September 1963, *Nair* 30548 (BSD). **Jammu & Kashmir**: without locality, 3048–3657 m, 24.8.1847, *Thomson* (CAL).

Notes: A very distinct species which can easily be identified by its cushion forming bushy and spiny habit, thick and hard stems, dense, stiff, persistent and spine-tiped rachis, deeply clift and very densely wooly calyx and included pods into the calyx. It is very common in Kaza-Rangrik areas (Spiti region) and its adjacent region of Kunawur (Kinnaur district) in Himachal Pradesh.

26. *Astragalus* sect. *Sesamei*

DC., Prodr. 2: 287. 1825; Podlech & Zarre, Tax. Rev. gen. *Astragalus* (Legum.) Old World 1: 63. 2013. –Type species: *Astragalus sesameus* L.

Annual herbs, stems well developed, hairy with basifixed hairs. Stipules adnate to petiole, free from each other or shortly connate behind stem. Leaves sessile or petioled. Inflorescence sessile or peduncled dense head. Calyx campanulate or shortly tubular. Keel petals shorter than wing petals. Pods sessile, erect or spreading, narrowly oblong to narrowly ovoid, dilated at base, bilocular or rarely semibilocular.

1. ***Astragalus tribuloides*** Delile, Descr. Egypte, Hist. Nat. 70. 1813; Boiss., Fl. Or. 2: 221. 1872; Baker in Hook.f., Fl. Brit. India 2: 122. 1876; Gontsch. & Popov in Komarov, Fl. U.R.S.S. 12: 31. 1946; Dhillon & Bajwa, Bull. Bot. Surv. India 11 (3&4): 237. 1969; Ali in Nasir & Ali, Fl. W. Pakistan 100: 146, f. 17 F. 1977; Sanjappa, Legum. India 96. 1992; Kumar & Sane, Legum. South Asia: Checkl. 243. 2003; Xu & Podlech in Wu *et al.*, Fl. China 10: 337, f. 355. 12–18, 2010; Podlech & Zarre, Tax. Rev. gen. *Astragalus* (Legum.) Old World 1: 93. 2013. –Type: Egypt, *Delile s. n.* (lecto MPU *n. v.*; iso P!: photo LWG!).

Plate 79

Plants annual herbs, prostrate, caespitose, several stems arising from the base, stems slender, terete, faintly striate, moderately to densely pilose with appressed or half appressed, forwardly oriented, mixed white and blackish-brown hairs. Stipules

2.5–3.5 × c. 1.5 mm, adnate to the petiole at base, ovate or narrowly triangular, acuminate at tip, glabrous or glabrescent on both sides, ciliate along margins and tip. Leaves 6–8 cm long, imparipinnately compound; petiole 1.5–1.8 cm long; rachis 4–6.5 cm long, hairy as stem; leaflets 6–8 pairs, 5–11 × 2–3 mm, subsessile, opposite or subopposite, sometimes alternate, oblong, lanceolate, narrowly lanceolate or oblanceolate, cuneate at base, entire along margins, acute at base, pilose on both sides with white, spreading hairs. Inflorescence axillary, sessile to shortly peduncled capitate racemes, 2–many-flowered; peduncle 2–2.5 mm long (up to 20 mm long in fruits), densely pilose with spreading, mixed white and blackish-brown hairs. Bracts c. 2 mm long, linear-ovate with ciliate margins. Flowers pinkish-white, pale pink or white, c. 6 mm long, subsessile; pedicel less than 1 mm long, pilose with spreading, mixed white and blackish-brown hairs. Calyx c. 4.3 mm long, campanulate, pilose with mixed white and blackish-brown hairs outside, glabrous inside, tube c. 2.3 mm long, lobes c. 2 mm long, linear, subequal, shorter on dorsal side, longer on ventral side. Standard c. 6 × 2 mm, oblong, retuse at apex ; wing petals 5–5.5 mm long, lamina 2– 2.5 mm long, linear-oblong, obtuse at tip, claw c. 3 mm long; keel petals c. 4.5 mm long, lamina c. 1.5 mm long, upper auricle poorly developed, claw c. 3 mm long. Ovary c. 3 mm long, sessile, hairy; style c. 1 mm long; stigma glabrous. Pods 6–10 mm long, sessile, arranged like star, narrowly ovoid, slightly arched, turgid, gibbous at base, acute at tip, grooved dorsally, bilocular, pubescent with mixed long and short white hairs, long hairs spreading and forwardly oriented. Seeds c. 1.2 × 1 mm, brown, smooth, glabrous.

Flowering & Fruiting: February–March.

Habitat: A weed of cultivated field in sandy soils.

Distribution: INDIA: Punjab, Haryana, Rajasthan; MIDDLE EAST; NORTH AFRICA; NORTHERN EURASIA; PAKISTAN; WEST ASIA.

Specimens examined: **Haryana**: Hissar, Talwandi, 2.3.1962, *Nair* 19861 (LWG, BSD); Hissar, Barwala, 13.1.1962, *Nair* 19275 (BSD); Hissar, Chowdhrawas, 28.2.1962, *Nair* 19396 (BSD). **Punjab**: Dabwali, 20.2.1963, *Nair* 26159 (BSD); Dabwali, 20.3.1963, *Nair*, 26149 (BSD); Bahuna, 11.2.1963, *Nair* 25985 (BSD); Firozpur dist, Lambi, 21.2.1963, *Nair* 26208, 26181 (BSD); Punjab, region temp., 307m, *T. Thomson s. n.* (BM).

Notes: Recorded here from Rajasthan after Dhillon & Bajwa (1969).

27. *Astragalus* sect. *Skythropos*

N. D. Simpson, Notes Roy. Bot. Gard. Edinburgh 8: 255. 1915; Podlech & Zarre, Tax. Rev. gen. *Astragalus* (Legum.) Old World 1: 295. 2013. –Type species: *Astragalus skythropos* Bunge.

Perennial herbs, with short to long stems, hairy with basifixed white or black hairs. Stipules free or lower ones connate behind stem, scale like and without leaves in basal part of stem. Racemes long peduncled, 1–many-flowered. Stigma glabrous or hairy. Pods stipitate, unilocular.

Key to the Species

1a. Bracteoles absent; flowers pink with purplish tinge; stigma glabrous; pods covered with black hairs ....1. ***A. kongrensis***

1b. Bracteoles sometimes present; flowers yellow; stigma sometimes minutely hairy; pods covered with white and black hairs ....2. ***A. yunnanensis***

1. ***Astragalus kongrensis*** Baker in Hook. f., Fl. Brit. India 2: 120. 1876; Sanjappa, Legum. India 90. 1992; Kumar & Sane, Legum. South Asia: Chekl. 232. 2003; L.B. Chaudhary, Taiwania 51 (3): 237, f. 1. 2006; Podlech & Zarre, Tax. Rev. gen. *Astragalus* (Legum.) Old World 1: 297. 2013. –Type: India, Sikkim, Kongra Lama, 15000 ft., 24.7.1849, *J. D. Hooker s. n.* (holo K!).

Plants rhizomatous subacaulescent herbs, stems c. 5 cm long, ascending to erect, often zig-zag like bent, pilose with long, white, spreading hairs. Stipules 8–10 × 3–4 mm, foliaceous, free from petiole, free from each other or connate behind the stem and before the petiole in lower leaves, covering the stem, ovate-lanceolate, ciliate along margins with long, white, spreading hairs. Leaves 8–12 cm long, imparipinnately compound; petiole c. 3 cm long, rachis and petiole thinly pilose with long, white, spreading hairs; leaflets 6–15 pairs, 5–7 × 3–5 mm, broadly elliptic (almost rounded) or broadly obovate, subtruncate, obtuse or emarginated at apex, rounded at apex, ciliate along margins, glabrescent to thinly pilose with long, white, spreading hairs below, glabrous above. Inflorescence c. 12 cm long, axillary, peduncled capitate raceme (head); raceme dense, 5–12-flowered; peduncle c. 10 cm long, pilose with long, mixed blackish-brown and white, spreading hairs. Bracts c. 5 × 1 mm, longer than pedicel, linear, ciliate, pilose outside with long, white, spreading hairs. Flowers 15– 20 mm long, pink with purple tinge, nodded; pedicel c. 2 mm long, densely pilose with long, blackish-brown, half spreading hairs. Calyx c. 7 mm long, tubular-campanulate, pilose with long, mixed blackish-brown and white, spreading hairs outside, tube c. 4.5–5 mm long, lobes 2.5–3 mm long, shorter than tube, linear-lanceolate. Standard 15–20 × 7–9 mm, obovate, emarginate at apex; wing petals c. 13 mm long, shorter than standard and keel petals, lamina 7–8 × 2–3 mm, oblong, upper auricle linear; keel petals equal to standard, lamina c. 8 × 4 mm, obovate, curved towards apex. Ovary c. 5 mm long, densely appressed hairy with c. 5 mm long stipe (as long as calyx tube); style 6–8 mm long, glabrous; stigma glabrous. Pods (after Podlech & Zarre, 2013) 12–15 × c. 8 mm, widely ellipsoid, with c. 2 mm long beak and persistent glabrous style, unilocular, densely hairy with spreading black hairs.

Flowering & Fruiting: June–August.

Habitat: River banks and open grassy hillsides. Altitudes 3800–4200 m.

Distribution: INDIA: Sikkim; BHUTAN; CHINA.

Specimens examined: **Sikkim**: Gochung, 4.7.1903, *Younghusband* 15 (CAL); Between Gochung-Talling, 21.7.1986, *Raju* 6159 (BSHC).

2. *Astragalus yunnanensis* Franch., Pl. Delav. 162. 1889; Wenninger, Mitt. Staatss. Munchen 30:146, f. 13 f-h, 25 o-p. 1991; Kumar & Sane, South Asia Legum.: Checkl. 244. 2003; L.B.Chaudhary & S.K.Srivast., Taiwania 52 (1): 30. 2007; Xu & Podlech in Wu *et. al.*, Fl. China 10: 354, f. 382. 1-22. 2010; Podlech & Zarre, Tax. Rev. gen. *Astragalus* (Legum.) Old World 1: 300. 2013. –Type: China, Yunnan, au pied du glacier de Likiang, 3800 m, 9.7.1884, *Delavay s. n.* (lecto P *n. v.*; iso K!, L *n. v.*, MSB *n. v.*).

A. fenzelianus E. Peter, Acta Horti. Gothob. 12: 54. 1938; Wenninger, Mitt. Bot. Staatssamml. Munchen 30: 134, f. 12 a-b. 1991; Kumar & Sane, South Asia Legum.: Checkl. 228. 2003; L.B. Chaudhary & S.K.Srivast., Taiwania 52 (1): 27. 2007. –Type: Tibet Or., Yeniu-schan ad lac. Kukunor, 22–23.7.1935, *Fenzel* 2343 (holo W *n. v.*).

Plants 15–25 cm long, subacaulescent to shortly caulescent; stems many, 1–10 cm long, ascending to erect, glabrous or sparsely covered with spreading white hairs in upper part, basal parts covered with connate stipules. Stipules 4–14 × 3–8 mm, lower ones widely ovate to elliptic, upper ones narrowly triangular, all free, sparsely hairy. Leaves 5–16 cm long, imparipinnately compound; petiole 2–4.5 cm long, like rachis glabrous or sparsely hairy; leaflets 5–15 pairs, 3–15 × 2.5–10 mm, ovate to elliptic, obtuse to acute at apex, sparsely to densely villous below, sometimes hairy only along midvein and margins, glabrous or sparsely to more rarely densely villous below, often with a distinct 1–2 mm long cusp. Inflorescence 5–13.5 cm long, raceme, densely 2–13-flowered; peduncle 4–11 cm long, densely white and blackish-brown hairy; floral rachis 1–2.5 cm long, elongating up to 7 cm at fruiting time. Bracts 4–10 × 1–3 mm, membranous, narrowly ovate to ovate, mostly white hairy. Bracteoles sometimes present at the base of calyx. Flowers pale yellow, sometimes slightly fading violet, nodded; pedicel 1–2 mm long, predominantly coverd with blackis-brown hairs. Calyx 8–14.5 mm, campanulate-tubular, tube 4–7 mm long, truncate to oblique at mouth, covered with white and black or only black hairs; teeth 4–7.5 mm long, nearly equal, narrowly triangular. Standard 16–21 × 7–13.5 mm, widely elliptic to widely ovate, emarginated at apex; wing petals equal to standard, lamina 10.5–15 × 3–4.5 mm, narrowly oblong to narrowly obovate, claw 8–11 mm long; keel petals 16.5–23 mm long, lamina 8–13 × 2.5–5.5 mm, curved on lower edge, obtuse at apex, claw 8–13 mm long. Staminal sheath nearly truncate at tip. Ovary with 4–5 mm long stipe, densely white and blackis-brown hairy; style glabrous; stigma occasionally minutely hairy. Pods 15–23 × 6–8 mm, narrowly ovoid to narrowly ellipsoid, with 4–8 mm long stipe and 2–3 mm long beak, densely covered with spreading white and black hairs.

Flowering & Fruiting: July–August.

Habitat: Altitude 3960–4365 m

Distribution: INDIA: Uttarakhand; CHINA; BHUTAN; NEPAL.

Specimens examined: **Uttarakhand**: Kumaon, Nipchang valley, Darma, 4365 m, 31. 8. 1884, Duthie 2814 (K).

28. *Astragalus* sect. *Uliginosi*

A. Gray, Proc. Amer. Acad. 6: 196. 1864; Podlech & Zarre, Tax. Rev. gen. *Astragalus* (Legum.) Old World 2: 1540. 2013. –Type species: *Astragalus uliginosus* L.

Perennial herbs, stems well developed, covered with medifixed hairs. Stipules free from petiole, connate opposite the petiole. Racemes long peduncled, loosely to densely many-flowered. Calyx tubular. Petals adnate to staminal sheath. Pods sessile, erect or nodded, incompletely or completely bilocular.

1. ***Astragalus peduncularis*** Royle ex Benth. in Royle, Illustr. Bot. Himal. Mount. 199. 1835; Baker in Hook. f., Fl. Brit. India 2:136. 1876; Gontsch. in Komarov, Fl. U.R.S.S. 12: 438. 1946; Ali, Biologia 7: 75. 1961 & in Nasir & Ali, Fl. W. Pakistan 100: 203. 1977; Sanjappa, Legum. India 92. 1992; Kumar & Sane, Legum. South Asia: Checkl. 236. 2003; Podlech & Zarre, Tax. Rev. gen. *Astragalus* (Legum.) Old World 2: 1545. 2013. –Type: India, Pungee in Kunawar, *Royle s. n.* (holo LIV *n. v.*; iso LE *n .v.*).

Plates 30, 80

Plants perennial herbs, prostrate to erect, up to 70 cm long, stems branched, terete, striate, sparsely pubescent below (older part), densely pubescent above (younger parts), hairs minute, completely appressed, medifixed, white. Stipules 2–5 × 1–2 mm, smaller in upper portion, free from petiole, connate up to the middle opposite the petiole, almost encircling the stem, lanceolate with more acuminate tip, pubescent outside, glabrous inside, hairs similar to stem but comparatively longer. Leaves 4–11.5 cm long, imparipinnately compound; petiole 5–20 mm long; rachis 3.5–9 cm long, channeled above; petiole & rachis pubescent like stem; leaflets 7–11 pairs, 6–12 × 1–4.5 mm, subsessile, opposite, subopposite or sometimes alternate, oblong, oblong-obovate (narrowly obovate) or ovate, cuneate at base, entire along margins, subacute, obtuse or subtruncate with very minute macro at apex, glabrous above, pubescent below, hairs as on stem. Inflorescence 13–22 cm long (c. 31 cm long in fruits) axillary, long peduncled raceme, many-flowered; peduncle 9–14 cm long, terete, striate, sparsely finely pubescent with medifixed, appressed, white hairs; floral rachis 5–8 cm long, moderately to densely pubescent, hairs similar to peduncle. Bracts 3–4 mm long, subpersistent , deflexed, linear-lanceolate, few minute glands present along margins, glabrous inside, pubescent outside. Bracteoles very minute, setaceous, pubescent, present at the base of calyx. Flowers 15–16 mm long, greenish-white or creamy-white, nodded, minutely pediceled; pedicel c. 1 mm long or less, pubescent. Calyx c. 8 mm long, persistent, membranous, oblique at base (the attachement of pedicel is ± lateral), tubular, tube c. 6.5 mm long, faintly veined, pubescent outside with mixed white & brown hairs, glabrous inside, lobes subequal, dorsal ones shorter and deltoid, c. 1 mm long, ventral ones c. 1.5 mm long, linear-lanceolate, all pubescent on both sides. Standard 15–16 mm long, lamina c. 8 × 6 mm, deflexed in upper part, obovate, deeply emarginate with a mucro at apex, claw c. 8 mm long; wing petals c. 13 mm long, shorter than standard and longer than keel petals, lamina c. 6.5 × 2.5 mm, oblong, subobtuse at apex, upper auricle c. 2 mm long, oblong with truncated apex, spurs faintly developed, claw c. 6.5 mm long; keel petals c. 11.5 mm long, curved in upper part, subacute at apex, upper auricle c. 1 mm long, claw c. 6 mm long. Staminal sheath c. 9 mm long, free, filaments unequal, 1.5–2.5 mm long; vexillary filament c. 11 mm long. Ovary c. 3 mm long, linear, sessile, glabrous, becoming pubescent at maturity, hairs similar to stem; style flat, glabrous, few hairs present in lower portion along margins; stigma capitate, glabrous. Pods 15–20 × 4–5 mm, nodded, oblong, sessile, turgid, grooved dorsally, narrowed at tip into a 2–3 mm long beak, with rugose and faintly veined surface, unilocular, dehiscent, pubescent with minute, appressed, medifixed, mixed white and brown hairs, 15–20-seeded. Seeds c. 2.5 × 1.5 mm, reniform, yellowidh-brown, turgid, smooth, glabrous.

Flowering & Fruiting: July–August.

Habitat: In dry hard rocky soils on flat grounds, hill slopes and along riversides. Altitudes 3000–4000 m.

Distribution: INDIA: Himachal Pradesh, Jammu & Kashmir, Uttarakhand; AFGHANISTAN; CHINA; KAZAKHSTAN; KIRGIZSTAN; PAKISTAN; TADZHIKISTAN; UZBEKISTAN.

Specimens examined: **Himachal Pradesh**: Lahul-Spiti: Gondla, 15.9.2001, *Chaudhary* 220668 (LWG); Darcha, 3500 m, 17.7.1979, *Aswal* 10321 (CDRI); Darcha, 3400 m, 16.9.1978, *Aswal* 9349 (CDRI); Darcha, 3500 m, 4.8.2003, *Chaudhary & Khan* 225018 (LWG); Darcha, 3231 m, N 32° 37.961′ E 77° 10.442′, 29.7.2006, *Chaudhary et al.* 229439 (LWG); Khurik, 3800 m, 6.8.1972, *Bhattacharyya* 4937 (BSD); Rangrik, 3600 m, *Bhattacharyya* 52022 (BSD); Guling-Mikkin, 15.8.1994, *Murti & Singh* (BSD); Thumla, 4200 m, 1.8.1972, *Bhattacharyya* 49200 (BSD); Between Keylong and Jispa, 3146 m, N 32° 34.166′ E 77°06.263′, 29.7.2006, *Chaudhary et al.* 229432 (LWG); Between Keylong and Jispa, 3146 m, N 32° 34.166′ E77°06.263′, 29.7.2006, *Chaudhary et al.* 229433 (LWG); Between Keylong to Jispa, 3. 8. 2014, *Chaudhary et al.* 259365 (LWG); Keylong, 3200 m, 7.6.1941, *Bor* 14694 (DD); 15 km before Kaza from Koksar side, 3700 m, 4.8.2002, *Chaudhary & Khan* 206769 (LWG); Lahaul, upper Bhagga valley, 2895–3962 m, June 1865, *Stoliczka s. n.* (CAL); Spiti, Peen valley, 3505–3657 m, *Stoliczka s. n.* (CAL); Kinnaur Dist., P.W.D. rest house, 2837 m, 7.6.1970, *Janardhanan* 47843 (BSD); Bashahr, Kailash range, 2743–3048 m, *Gorrie* (DD); Pangi,. *Stoliczka s. n.* (CAL). **Jammu & Kashmir**: Gilgit, March 1885, *Giles* 210 (DD); Baltistan, Turmih Nallah, 2743 m, 26. 5. 1928, *Ludlow* 314 (BM); Ladakh, Chilung, 3353 m, July 1905, *Meebold* 1382 (CAL); Ladakh, Suru valley, Panikhar, 3300 m, 24.8.1970, *Wadhwa* 59874 (BSD); Suru valley, near Panikhar village, 3263 m, N 34° 06.722′ E 75° 55.853′, 4.8.2011, *Chaudhary et al.* 263867 (LWG); Shingo valley, 3048–3353 m, 7.7.1892, *Duthie* 11886 (BM); Aster valley, Mushkin forest, 2743–3048 m, 27.7.1892, *Duthie* 12271 (CAL). **Uttarakhand**: Uttarkashi, Nago-Judung, 4000 m, August 1988, *Naithani* 86850 (BSD); Kumaon, 1831, without collector and collection number, acc. no. 114147 (CAL).

Notes: The pods are hairy in this species, however, Baker (1876) has observed glabrous pods. Podlech & Zarre (2013) have mentioned that the pods become glabrous with age. They have observed incompletely bilocular pods in this species, while in our specimens the pods are unilocular.

Excluded Species

1. ***Astragalus alitschuri*** B. Fedtsch., Trudy Imp. S.- Petersburgsk. Bot. Sada 21: 317. 1903; Ali, Phyton 12; 180. 1967 & in Nasir & Ali, Fl. W. Pakistan 100: 210. 1977; Kumar & Sane, Legum. South Asia: Checkl. 222. 2003; L.B. Chaudhary & S.K.Srivast., Taiwania 52 (1): 26, f. 1. 2007; Podlech & Zarre, Tax. Rev. gen. *Astragalus* (Legum.) Old World 3: 1852. 2013.

   Distribution: PAKISTAN, TADZHIKISTAN.

2. ***Astragalus charguschanus*** Freyn , Bull. Herb. Boiss., sér. 2, 4: 764.1905; Kumar & Sane, Legum. South Asia: Checkl. 222. 2003; L.B.Chaudhary & S.K.Srivast., Taiwania 52 (1): 27. 2007. *A. staintonianus* Ali, Candollea 17: 87. t. 13, 16. 1960; Biologia 7: 57. 1961; Phyton 11: 135. 1966 & in Nasir & Ali, Fl. W. Pakistan 100: 170. 1977; Sanjappa, Legum. India 94. 1992; Podlech & Zarre, Tax. Rev. gen. *Astragalus* (Legum.) Old World 1: 532. 2013.

   Distribution: AFGHANISTAN, PAKISTAN, TADZHIKISTAN.

3. ***Astragalus contortuplicatus*** L., Sp. Pl.: 758. 1753; Baker in Hook.f., Fl. Brit. India 2:122. 1876; Cooke, Fl. Bombay Pres. 1: 330. 1903; Ali, Biologia 7: 26. 1961 & in Nasir & Ali, Fl. W. Pakistan 100: 149, f. 17 J. 1977; Kumar & Sane, Legum. South Asia: A Checkl. 226. 2003; L.B.Chaudhary & S.K.Srivast., Taiwania 52 (1): 27. 2007; Podlech & Zarre, Tax. Rev. gen. *Astragalus* (Legum.) Old World 1: 153. 2013.

   Distribution: CAUCASUS, CENTRAL ASIA, CENTRAL EUROPE, EASTERN SIBERIA, PAKISTAN (PUNJAB, SIND, BALUCHISTAN), RUSSIA.

4. ***Astragalus flemingii*** Ali, Kew Bull. 1958: 311, t. 2 & 3. 1958; Biologia 7: 56. 1961 & in Nasir & Ali, Fl. W. Pakistan 100: 168, f. 24 A - D. 1977; Podlech, Mitt. Staatss. Munchen 25: 660. 1988; Sanjappa, Legum. India 87. 1992; Kumar & Sane, Legum. South Asia: A Checkl. 229. 2003; L.B.Chaudhary & S.K.Srivast., Taiwania 52 (1): 28, f. 3. 2007; Podlech & Zarre, Tax. Rev. gen. *Astragalus* (Legum.) Old World 1: 672. 2013. *A. bakuensis* auct. non Bunge 1868; Baker in Hook.f., Fl. Brit. India 2: 132. 1876.

   Distribution: AFGHANISTAN, PAKISTAN (PUNJAB).

5. ***Astragalus iodotropis*** Boiss. & Hohen., Boiss., Diagn. pl. orient., ser. 1, 9: 100. 1849; Srivastava *et. al.*, Indian J. Forestry 7: 335. 1984; Kumar & Sane, legume. South Asia: Checkl. 231: 2003; L.B.Chaudhary & S.K.Srivast., Taiwania 52 (1): 28. 2007; Podlech & Zarre, Tax. Rev. gen. *Astragalus* (Legum.) Old World 2: 1353. 2013.

Distribution: NORTHERN IRAN- ENDEMIC.

6. ***Astragalus pyrrhotrichus*** Boiss., Diagn. pl. orient., ser. 1(9): 73. 1849; Baker in Hook.f., Fl. Brit. India 2:133. 1876; Blatter, Beautf. Fl. Kashmir 92. 1928; Ali, Biologia 7: 53. 1961 & in Nasir & Ali, Fl. W. Pakistan 100: 168. 1977; Podlech, Mitt. Bot. Staatss. Munchen 25: 451. 1988; Sanjappa, Legum. India 93. 1992; Kumar & Sane, Legum. South Asia: Checkl. 238. 2003; L.B.Chaudhary & S.K.Srivast., Taiwania 52 (1): 29, f. 7. 2007; Podlech & Zarre, Tax. Rev. gen. *Astragalus* (Legum.) Old World 1: 617. 2013.

   Distribution: AFGHANISTAN, PAKISTAN.

7. ***Astragalus stocksii*** Bunge, Mém. Acad. Imp. Sci. Saint Pétersbourg 11 (16): 6. 1868 & 15 (1): 4. 1869; Cooke, Fl. Bombay Pres. 1: 330. 1903; Ali, Biologia 7: 72. 1961 & in Nasir & Ali, Fl. W. Pakistan 100: 203, f. 27 K–Q . 1977; Kumar & Sane, Legum. South Asia: A Checkl. 241. 2003 ; L.B.Chaudhary & S.K.Srivast., Taiwania 52 (1): 30. 2007; Podlech & Zarre, Tax. Rev. gen. *Astragalus* (Legum.) Old World 2: 1539. 2013.

   Distribution: AFGHANISTAN, IRAN, PAKISTAN.

8. ***Astragalus subumbellatus*** Klotzsch in Klotzsch & Garcke, Bot. Ergebn. Reise Wad. 159, t. 3. 1862; Baker in Hook.f., Fl. Brit. India 2:119. 1876, p.p.; Ali, Kew Bull. 1958: 305. 1958; Biologia 7: 17. 1961 & in Nasir & Ali, Fl. W. Pakistan 100: 132. 1977. L.B.Chaudhary & S.K.Srivast., Taiwania 52 (1): 29. 2007; Podlech & Zarre, Tax. Rev. gen. *Astragalus* (Legum.) Old World 2: 930. 2013. *A. hippocrepidis* Benth. ex Bunge., Astrag. 1: 5. 1868. *A. camporum* Benth. ex Boiss., Fl. Or. 2: 231. 1872.

   Distribution: AFGHANISTAN, PAKISTAN.

9. ***Astragalus zacharensis*** Bunge, Mém. Acad. Imp. Sci. Saint Pétersbourg 11 (16): 23. 1868; Stewart, Fl. W. Pakistan 398. 1972; L.B. Chaudhary & S.K.Srivast., Taiwania 52 (1): 31 . 2007; Podlech & Zarre, Tax. Rev. gen. *Astragalus* (Legum.) Old World 2: 1028. 2013.

   Distribution: CHINA.

Phyllolobium

Fisch., Sprengel, Novi Provent. 33. 1818; Zhang & Podlech, Feddes Repert. 117 (1-2): 41. 2006. –Type species: *Phyllolobium chinense* Fisch.

Astragalus sect. *Phyllolobium* (Fisch.) Bunge, Mém. Acad. Imp. Sci. Saint Pétersbourg 11 (16): 3. 1868.

A. sect. *Chlorostachys* subsect. *Phyllolobium* (Fisch.) Podlech in Rech.f., Fl. Iranica 174: 79. 1999.

A. subgen. *Trichostylus* Baker in Hook. f. Fl. Brit. India 2: 118. 1876. –Type species: *Astragalus heydei* Baker

Perennial herbs, generally with well develoved stems, with basifixed hairs. Stipules free from petiole and from each other, occasionally shortly connate opposite the petiole. Leaves stipulate; leaflets without stipels. Inflorescence axillary, peduncled raceme. Flowers with persistent Bracts. Bracteoles present or absent. Calyx campanulate. Wing and keel petals not interlocking due to absence of bosses on the keel petals and matching sockets on the wing petals. Stamens diadelphous (9 + 1) or monadelphous; vexillary stamen free. Style bearded at and near the apex on the inner side just below glabrous stigma. Pods unilocular or partially to completely bilocular. Seeds reniform.

23 species; chiefly occurs in the Southern-Western high mountains of China (21 species), only few extend towards Bhutan, Nepal, India,Pakistan and Tadzhikistan; 4 species in India.

Key to the Species

1a. Stamens closed monadelphous .... **1. *P. donianum***

1b. Stamens diadelphous (9+1) 2

2a. Plants 4–5 cm long; bracteoles absent; calyx c. 4 mm long; keel petals slightly shorter than wing petals .... **2. *P. heydei***

2b. Plants up to 30 cm long; bracteoles present; calyx 7–10 mm long; keel petals longer than wing petals 3

3a. Stems rooting at nodes in lower portion, glabrous to glabrescent; calyx tube longer than lobes, lobes oblong-ovate .... **3. *P. lachungense***

3b. Stems not rooting at nodes, densely tomentose; calyx tube more or less equal to lobes, lobes linear .... **4. *P. tribulifolium***

1. *Phyllolobium donianum* (DC.) M. L. Zhang & Podlech, Feddes Repert. 117: 49. 2006.

Astragalus donianus DC., Prod. 2: 283. 1825; Baker in Hook. f., Fl. Brit. India 2: 119. 1876; Chater in Hara *et al.*, Enum. Fl. Pl. Nepal 2: 106. 1979; Grierson & Long in Grierson & Long, Fl. Bhutan 3: 719. 1987; Sanjappa, Legum. India 87. 1992;

Kumar & Sane, Legum. South Asia: Checkl. 227. 2003. –Type: India, Emodo ad Gossainthan, *Wallich* Cat. 5927 (holo K!, iso BR *n. v.*, CGE *n. v.*, LIV *n. v.*).

A. macrorhizus D. Don, Fl. Nepal. 245. 1825, non Cav. 1793.

A. pycnorhizus Wall. ex Benth. in Royle, Illustr. Bot. Himal. Mount. 199. 1835; Baker in Hook. f., Fl. Brit. India 2: 119. 1876; Ali, Biologia 7: 10. 1961. –Type: that of *A. donianus* DC.

Plates 31, 81

Plants prostrate with thick woody rootstock or fibrous roots, profusely caespitose, stems 10–40 cm long, reddish, slender, glabrous to thinly pubescent with minute, white, flat, appressed hairs. Stipules c. 2 x 1–1.5 mm, free or slightly adnate to petiole, connate opposite the petiole in lower portion of the plant and completely encircling the stem, membranous, triangular or ovate-lanceolate, acute to subacuminate at apex, irregularly dentate along margins with glandular structure. Leaves 16–35 mm long, imparipinnately compound; petiole 0–5 mm long; rachis and petiole pubescent with white, appressed, flat hairs; leaflets 5–8 pairs, 3–6 x 1.5–3.5 mm, alternate to opposite, oblong, oblong-obovate or oblong-elliptic, cuneate or subobtuse at base, emarginate at apex, glabrous above, pubescent with white, flat, appressed to half spreading hairs below. Inflorescence axillary, peduncled 1–2-flowered raceme; peduncle 1.7–5 cm long, equal to or longer than subtending leaf, pubescent with minute, white or mixed white and brown, appressed hairs in upper portion,. Bracts 1.5–2 x c. 1.5 mm, membranous, persistent, encircling the pedicel, broadly ovate, acute to subcuminate at apex, glabrous, dentate along margins with some glandular structure. Bracteoles present at the tip of pedicel, 1–1.5 mm long, ovate-lanceolate, in other details similar to bract. Flowers 15–18 mm long, dark brownish-purple, blackis-brown or dark maroon; pedicel c. 1.5 mm long, densely pubescent with flat, brown or black, appressed hairs. Calyx 5–7 mm long, tinged with light brown colour, persistent, campanulate, pubescent with brown, flat, appressed hairs outside, tube 2.5–4 mm long, lobes c. 2.5–3 x c. 1 mm, equal to tube, ovate-lanceolate with long acuminate tip, densely hairy inside with black or rarely white hairs, glands present along margins. Standard 15–18 x 11–13 mm, longer than wing and keel petals, backwardly deflexed, suborbicular, emarginate at apex, claw c. 3 m long; wing petals c. 14 mm long, slightly shorter than or sometimes more or less equal to keel petals, lamina c. 9 x 4 mm, oblong-ovate, obtuse at apex with a slight notch near the apex on ventral margin, upper auricle c. 1.5 x 1.5 mm, claw c. 5 mm long; keel petals c. 14 mm long, lamina c. 9 x 5 mm, narrowly obovate with incurved towards apex, obtuse at apex, upper auricle c. 2 x 2 mm, claw c. 5 mm long. Stamens monadelpous, closed, split on dorsal side after anthesis, staminal sheath c. 10 mm long, obtuse at apex,free filaments 1.5–2.5 mm long, alternately shorter and longer. Ovary c. 9 mm long, with c. 3 mm long stipe, minutely pubescent with white hairs; style c. 4 mm long, bearded at and near the apex on the inner side; stigma penicillate. Pods 9–18 x 4–9 mm, partially tinged with light brown colour, oblong-ellipsoid, narrowed at apex into a 7–8 mm long beak, turgid with intruded dorsal suture, unilocular, shortly stipitate, pubescent with very minute, brown or brownish-black, appressed hairs, stipe c. 1 mm long.

Flowering & Fruiting: July–September.

Habitat: Common (in Sikkim) in open grassy hill slopes, along agriculture fields, on rocky flat lands and river sides in sandy as well as hard gravelly soils. Altitudes (2700–) 3000–4000 (–4600) m.

Distribution: INDIA: Arunachal Pradesh, Sikkim, Uttarakhand; BHUTAN; CHINA; NEPAL.

Specimens examined: **Arunachal Pradesh:** Pukdua-Gambesi, 13500–14000–9000 ft., 31.05.1958, *Rao* 17202 (CAL). **Sikkim:** Zemu valley, 11800 ft., 11.7.1909, *Smith & Cave* 1207 (CAL); Zemu valley, 9600 ft., 9.7.1909, *Smith & Cave* 1047 (CAL); Chola Range, Barfonchar, 11000 ft., 26.9.1892, *Gammie* 1301 (CAL); Thaling, 13000 ft., 1885, *King's collector* (CAL); Cheongtong, 5000 ft., 28.7.1892, *Gammie* 301 (CAL); Without locality, 8000–12,000 ft., Date?, *Hooker* (CAL); Samdong, 1901, *Prain* 221 (CAL); Llonak, 13000 ft., 30.7.1909, *Smith & Cave* 2774 (CAL); Lachen, 8000 ft., September–Ocober 1909, *Ribu & Rhomoo* 3049 (CAL, LBG); Lachen, Zema, near water fall, 3387 m, N27° 49.229′ E 88° 32.959′, 4.9.2015, *Chaudhary & Kushwaha* 259382 (LWG); Lachen, near Zema, N27°47.718′E88°33.147′, 3215 m, 4. 9. 2015, *Chaudhary & Kushwaha* 259379, 259378 (LWG); Between Lachen and Thangu, Kelep, 3800 m, 12.8.2004, *Chaudhary* 225064 (LWG); Between Lachen and Thangu, 3300 m, 12.8.2004, *Chaudhary* 225062 (LWG); Thangu, 4000 m, 12.8.2004, *Chaudhary* 225067 (LWG); upwards Thangu, along Lashachu, 3400 ft. 22.7.1988, *Raju & Singh* 8890 (BSHC); Lachen proper, near bridge, 2800 m, 16.8.2004, *Chaudhary* 225086 (LWG); Between Lachung and Yungthang, 15 km away from Yungthang, 3300 m, 15.8.2004, *Chaudhary* 225080 (LWG); North of Lachung, Between Phumi-Yacke, 3310 m, 14.7.1996, *Sinha & Long* 17839 (BSHC); 2 km away from Yungthang, near Garam Pani, 3800 m, 15.8.2004, *Chaudhary* 225084 (LWG);Yumthang-Shivamandir, 3600 m, 29.7.1990, *Kumar & Singh* 11611 (BSHC); Yumthang, 13000 ft., 23.10.1924, *Rhomoo Lepcha* 2276 (LBG); **Uttarakhand**: Kumaon; Bhalek in Kali Valley, Byans, 11000–12,000 ft., 23.7.1886, *Duthie* 5482 (CAL).

Notes: Baker (1876) has treated *A. pycnorhizus* and *A. donianus* as distinct species. However, he could not examine the specimens of *A. donianus* and included it based on De Candolle (1825) and Bunge (1868). As per his description these two species differ mainly in calyx teeth. In *A. pycnorhizus* the calyx teeth are lanceolate and equal to tube while in *A. donianus* the teeth are half of the tube. In recent work (Zhang & Podlech, 2006), *A. pycnorhizus* has been treated as synonym of *A. donianus*.

2. *Phyllolobium heydei* (Baker) M. L. Zhang & Podlech, Feddes Repert. 117: 53. 2006.

Astragalus heydei Baker in Hook.f., Fl. Brit. India 2:118. 1876; Ali, Biologia 7: 12. 1961 & in Nasir & Ali, Fl. W. Pakistan 100: 126. 1977; Sanjappa, Legum. India 89. 1992; Kumar & Sane, Legum. South Asia: Checkl. 230. 2003. –Type: West Tibet, by the Tso–kar sea, between Rupschu and Korsog, October1842, *Heyde s. n.* (lecto K *n. v.*: photo CAL!, LWG !; iso CAL!: photo LWG!).

A. hendersonii Baker in Hook. f., Fl. Brit. India 2: 120. 1876. –Type: Tibet, upper part of Karakash valley, *Henderson s. n.* (holo K *n. v.*).

A. heydei var. *hendersonii* (Baker) Ohaba *et al.*, J. Jap Bot. 70: 17. 1995.

Plate 82

Plants small herbs with subterranean caudex covered with leafless stipules, 4–5 cm long, stems glabrous, internodes longer than stipules and covered with some membranous stipules in lower portion. Stipules c. 2 x 2 mm, free (slightly connate at base in lower portion, opposite the petiole), encircling the stem, thick, deltoid or broadly ovate with obtuse tip, glabrous or ciliate along margins in young condition. Leaves imparipinnately compound, c. 16 mm long; petiole c. 5 mm long, pubescent with white, appressed hairs; rachis c. 8 mm long, hairy as petiole; leaflets c. 6 pairs, c. 4 x 3 mm, thick, opposite, crowded, sessile, conduplicate, pilose with appressed, white hairs below, glabrous above. Inflorescence peduncled capitate raceme, 2–4-flowered; peduncle c. 15 mm long, thinly pubescent with appressed, minute, white hairs. Bracts c. 2 mm long, longer than pedicel, deltoid or ovate with subobtuse tip, thinly pilose outside with white, spreading hairs, ciliate along margins. Flowers c.10 mm long, rose pink; pedicel c. 1 mm long, pubescent with blackish-brown hairs. Calyx c. 4 mm long, campanulate, tube c. 2.5 mm long, pilose with appressed, mixed brown and white hairs outside, glabrous inside, teeth c. 1.5 mm long, outside hairy as tube, inside hairy only with brown hairs. Standard 9–10 mm long, rounded, emarginate at tip, reflexed; wing petals equal to or slightly shorter than standard, lamina c. 6 x 3 mm, obtuse at tip, claw c. 3 mm long; keel petals c. 8.5 mm long, slightly shorter than wing petals . Pods (after Zhang & Podlech, 2006) 15–20 mm long, subsessile, ellipsoid, without distinct beak, unilocular, densely hairy with appressed to ascending white or black and white hairs.

Flowering & Fruiting: July–Sept.

Distribution: INDIA: Jammu & Kashmir; CHINA; NEPAL.

Specimens examined: **Jammu & Kashmir:** Kyensa La Rupshu, 5400 m, *Koelz* 2229 (BM).

Notes: It is a very rare species in India. Except the Type specimen no other specimens are available in the Indian herbaria.

3. ***Phyllolobium lachungense*** (L.B.Chaudhary) Podlech, Feddes Repert. 120: 58. 2008.

Astragalus lachungensis L.B.Chaudhary, J. Jap Bot. 81 (3): 168, f. 1 & 2. 2006. –Type: India, Sikkim, Lachung, between Lachung and Yumthang, c. 15 km beforeYumthang, 3300 m, 15.8.2004, *Chaudhary* 225079 (holo LWG!; iso CAL!).

Plants prostrate herbs, up to 30 cm long, densely caespitose, stems slender, partially reddish, rooting at nodes in lower portion, glabrous to glabrescent, internodes longer than stipules. Stipules 1–2 x 1–2 mm, free, encircling the stem, ovate, acute or subobtuse at apex, irregularly dentate along margins with glandular structure, glabrous. Leaves 3–7 cm long, imparipinnately compound; petiole c. 15 mm long; rachis and petiole partially reddish, glabrous to moderately white, appressed hairy; leaflets 4–8 pairs, 5 – 8 x 2.5–4.5 mm, oblong-obovate, oblong-ovate or oblong-elliptic, cuneate or subobtuse at base, retuse or emarginated at apex, glabrous above, pubescent with white, appressed, flat hairs below. Inflorescence axillary, peduncled, 1-flowered; peduncle 16–35 mm long, shorter than subtending leaf, partially reddish, pubescent with minute, white, appressed hairs. Bracts 1.5–2.5 x 1–2 mm, equal to or slightly shorter than pedicel, submembranous, persistent, encircle the pedicel, broadly ovate, dentate along margins with some glandular structure, acute at apex, glabrous. Bracteoles 2, minute, situated at the tip of pedicel, in other details similar to Bracts. Flowers 16–18 mm long, bluish-violet; pedicel c. 2–2.5 mm long, pubescent with mixed black and white or light brown, appressed hairs. Calyx 7–10 mm long, persistent, tubular-campanulate, partially reddish, pubescent with minute, light brown, appressed hairs outside, glabrous inside except teeth, tube 5–7 mm long, teeth 2–3 x c. 2 mm, shorter than tube, oblong-ovate, acute at apex, densely black hairy inside. Standard 16–18 mm long, backwardly deflexed, lamina c.10 x 10 mm, orbicular, emarginate at apex, claw c. 6 mm long; wing petals 12–15 mm long, shorter than standard and keel petals, lamina c. 8.5 x 3 mm, oblong-ovate, obtuse at apex, claw c. 5.5 mm long; keel petals 14–18 mm long, shorter than or subequal to standard, lamina c. 9 x 5 mm, oblong-obovate, obtuse at apex, claw c. 7 mm long. Stamens diadelphous, (9 + 1); vexillary filament c. 12.5 mm long, free; staminal sheath c. 12 mm long, obtuse at apex, free filaments c. 1.5 - 2 mm long. Ovary 6–7 mm long, with 2.5–3 mm long stipe, finely pubescent with appressed, mixed white and brown hairs; style 6–7 mm long, incurved towards apex, bearded near the apex on the inner side; stigma capitate, penicillate. Pods 22–24 x 12–13 mm, stipitate (stipe c. 3 mm long), oblong, abruptly narrowed at both ends, inflated, unilocular, pubescent with minute, brown, appressed hairs. Seeds c. 1.5 x 1 mm, reniform, subturgid, smooth, black.

Flowering & Fruiting: August–October.

Distribution: INDIA: Sikkim. Endemic & rare.

Habitat: On mountain slopes among herbs on moist soil. Altitudes 3300 m.

Specimens examined: **Sikkim:** Lachung, between Lachung and Yumthang, c. 15 km beforeYumthang, 3300 m, 15.8.2004, *Chaudhary* 225079 (LWG); without locality, 29.10.1884, *King's collector s. n.*, acc. nos. 1135554, 113555, 113556, 113557, 113558, 113559, 113560 (CAL).

4. ***Phyllolobium tribulifolium*** (Bunge) M. L. Zhang & Podlech, Feddes Repert. 117: 60. 2006.

Astragalus tribulifolius Benth. ex Bunge, Mém. Acad. Imp. Sci. Saint Pétersbourg 11 (16): 4. 1868; Baker in Hook. f., Fl. Brit. India 2: 120. 1876; Ali, Biologia 7: 10. 1961 & in Nasir & Ali, Fl. W. Pakistan 100: 125. 1977; Sanjappa, Legum. India 96. 1992; Kumar & Sane, Legum. South Asia: Checkl. 243. 2003; L.B.Chaudhary & Z.H.Khan, Rheedea 13: 73, f. 1. 2003. –Type: W. Tibet, 16000 ft., *Thomson s. n.* (holo K!).

A. tribulifolius var. *pauciflorus* Marq., J. Linn. Soc., Bot. 48: 171. 1929. –Type:Tsela Dzong, Tibet, 2700–3000 m, 22.5.1924, *Kingdon Ward* 5692 (holo K *n. v*).

Plate 83

Plants prostrate, ascending or erect, stems several, 6–20 cm long, densely tomentose with white, spreading hairs. Stipules 3–4 x c. 1 mm, free, persistent, triangular-lanceolate with more acuminate tip, ciliate along margins, sparsely pilose with white, spreading hairs outside, glabrous inside. Leaves 3.5–5 cm long, imparipinnately compound; petiole minute to 6 mm long, hairy as stem; rachis 2.2–4 cm long, hairy as stem; leaflets 5–7 pairs, 8–12 x 4–6.5 mm, subsessile, alternate to subopposite, usually obovate, sometimes oblong-obovate or obovate-elliptic, cuneate at base, entire along margins, retuse, emarginate, truncate or obtuse with a minute mucro at apex, pilose below with white, spreading hairs, glabrous to glabrescent above. Inflorescence c. 5 cm long, axillary, peduncled capitate raceme, 1–4-flowered; peduncle 15–25 mm long, more or less equal to subtending leaves, tomentose with mixed black and white hairs. Bracts c. 4 mm long, longer than pedicel, linear-lanceolate, hairy outside with spreading, white or mixed black & white hairs. Bracteoles 2, 2–4 mm long, situated at the base of calyx, hairy outside. Flowers 14–16 mm long, violet or creamy; pedicel c. 2 mm long, densely hairy with spreading, black hairs. Calyx 8–9.5 mm long, campanulate, pilose outside with spreading mixed black & white hairs, throat black hairy inside, tube c.4 mm long, lobes unequal, linear, dorsal ones c. 3.5 mm long, ventral ones 5–5.5 mm long. Standard 14–16 mm long, deflexed, lamina 10–12 x c. 11 mm, rounded, emarginate at apex, claw c. 4 mm long; wing petals 12–13 mm long, shorter than standard & keel petals, lamina c. 7.5 x 3.5 mm, upper auricle c. 1 mm long, hooked, claw c. 4.5 mm long; keel petals 14–15 mm long, equal to or slightly shorter than standard, lamina c. 9 x 5 mm, upper auricle c. 1.5 x 1 mm; claw c. 5 mm long. Stamens diadelphous, (9+1); vexillary filament c.11 mm long; staminal sheath c. 10 mm long, obtuse at apex, free filaments alternately shorter and longer, 1.5–2.5 mm long. Ovary c. 6 x 1 mm, stipitate, many-ovuled, stipe c. 3 mm long, shorter than calyx; style c. 4 mm long; stigma capitate, penicillate. Pods c. 15 x 7 mm, oblong-ellipsoid , turgid, grooved dorsally, stipitate, narrowed into a 1–2 mm long beak at tip, unilocular, pubescent with white, spreading hairs, stipe c. 2 mm long.

Flowering & Fruiting: July–September (–October).

Habitat: Growing on rocks on a very steep pine-bracken slopes in open area at 2770–5400 m altitudes.

Distribution: INDIA: Arunachal Pradesh , Jammu & Kashmir; CHINA, NEPAL, PAKISTAN(?).

Specimens examined: **Arunachal Pradesh** ('Assam'): Walong, 2400 m, 10.10.1950, *Kingdon-Ward* 20242 (BM). **Jammu & Kashmir:** Ladakh: Lorikung, 3900 m, 18.3.1980, *Bhattacharyya* 71729 (BSD); Muklj-Lukung, 3975–4175 m, 29. 08. 1976, *B. M. Wadhwa* 60156 (BSD).

Bibliography

Adiguzel, A., Agar, G., Baris, O., Gulluce, M., Sahin, F. and Sengul, M. 2006. RAPD and FAME analyses of *Astragalus* species growing in eastern Anatolia region of Turkey. *Biochem. Syst. Ecol.* **34**: 424–432.

Agerer-Kirchhoff, C. 1976. Revision von *Astragalus* L. sect. *Astragalus* (Leguminosae). *Boissier* **25:** 1–197.

Agerer-Kirchhoff, C. and Agerer, R. 1977. Eine neue Section der Gattung *Astragalus* L.: *Laxiflori* Agerer-Kirchhoff. *Mitt. Bot. Staatss. Munchen* **13:** 203–234.

Aitchison, J.E.T. 1881–1882. *On the flora of the Khurram valley & C. Afghanistan* - Part I & II. *J. Linn. Soc. Bot.* **18:** 1–122, 139–200.

Aitchison, J.E.T. 1888. The Botany of Afghan Delimitation Commission. *Trans. Linn. Soc. London Bot.* **3:** 1–139.

Ali, S.I. 1958. Notes on the genus *Astragalus* Linn. from W. Pakistan and N.W. Himalayas. *Kew Bull.* **1958**: 303–318.

Ali, S.I. 1960. Three new species of *Astragalus* L. from West Pakistan. *Candollea* **17:** 87–95.

Ali, S.I. 1961. Revision of the genus *Astragalus* L. from W. Pakistan and N.W. Himalayas. *Biologia* 7 (1 & 2): 1–92.

Ali, S.I. 1965. Contribution to the genus *Astragalus* from West Pakistan I. The subgenera *Pogonophace* and *Astragalus*. *Bot. Notiser* **118** (1): 87–96.

Ali, S.I. 1966. Contribution to the genus *Astragalus* from West Pakistan II. The subgenera *Caprinus* and *Cercidothrix*. *Phyton* (Horn) **11**: 133–140.

Ali, S.I. 1967. Contribution to the genus *Astragalus* from West Pakistan III. Subgenus *Tragacantha*. *Univ. Stud.* (Karachi) **3** (3): 40–45.

Ali, S.I. 1967. Contribution to the genus *Astragalus* from West Pakistan IV. Further additions to the subgenera *Pogonophace* and *Astragalus*. *Sind Univ. Sci. Res. J., Sci. Ser.* **2** (2): 179–183.

Ali, S.I. 1967. Contribution to the genus *Astragalus* from West Pakistan V. Further additions to the subgenus *Caprinus* and *Cercidothrix*. *Phyton* (Horn) **12**: 177–181.

Ali, S.I. 1973. Contributions to the genus *Astragalus* (Leguminosae) from Pakistan VI. Further additions and corrections. *Kew Bull.* **28** (2): 297–304.

Ali, S.I. 1977. Papilionaceae. In: Nasir, E. and Ali, S.I. (eds.), *Flora of West Pakistan* **100:** 1–389. Karachi, Pakistan.

Anand, K.K., Srivastava, R.K. and Chaudhary, L.B. 2010. Analysis of genetic diversity in *Astragalus rhizanthus* Benth. ssp. *rhizanthus* var. *rhizanthus* (Fabaceae) using molecular markers from India. *J. Botany* vol. **2010**, Article ID 701975, 9 pages, doi: 10.1155/2010/701975.

Anand, K.K., Srivastava, R.K., Chaudhary, L.B. and Singh, A.K. 2010. Delimitation of very closely allied Himalayan species of the *Astragalus rhizanthus* complex (Fabaceae) using molecular markers RAPD, ISSR and DAMD. *Taiwania* **55** (3): 197–207.

Anand, K.K., Jena, S.N. and Chaudhary, L.B. 2012. Reestablishment of the taxonomic status of *Astragalus malacophyllus* Benth. ex Bunge (Fabaceae) based on morphology and nrDNA Internal Transcribed Spacer sequence data. *Proc. Indian natn. Sci. Acad.* **78** (2): 171–187. 2012.

Anonymous. 1985. *The Wealth of India* (Revised) **1** (A): 476–477. Publication and Information Directorate, CSIR, New Delhi.

Aswal, B.S. 1994. *Flora of Lahul-Spiti District, Himachal Pradesh.* Bishen Singh & Mahendra Pal Singh, Dehra Dun.

Aswal, B.S. and Mehrotra, B.N. 1980. Contribution to the flora of Lahul valley (North-West Himalaya)-II some new plant records. *J. Econ. Tax. Bot.* **1**: 115–117.

Baker, J.G. 1876. Leguminosae. In: Hooker, J. D, *The Flora of British India* **2**: 56–306. Rev. & Co. Kent, London.

Barneby, R.C. 1964. Atlas of North American *Astragalus*. *Mem. New York Bot. Gard.* **13**: 1–1188.

Barneby, R.C. 1971. Nomenclatural Notes on Old World *Astragalus*. *Feddes Repert.* **82**: 579–580.

Baum, H. 1948. Vergleichend morphologische, anatomische und entwicklungsgeschichtliche Untersuchungen uber die *Astragalus*-Frucht. *Ann. Naturhist. Mus. Wien* **56:** 246–261.

Becht, R. 1978. Revision der Section *Alopecuroidei* DC. der Gattung *Astragalus* L. *Phanerog. Monograph* **10**: 1–227. Vaduz.

Benthum, G. 1935. Himalayan Leguminosae of European and Siberian forms. In: Royle, J. F., *Illustrations of the Botany of the Himalayan Mountains and of the Flora of Cashmere*. **1**: 197– 201; **2**: tab. 36.

Blatter, E. 1914–1916. Flora of Aden. *Rec. Bot. Survey India* **7**: 1– 418.

Blatter, E. 1921. Flora Arabica. *Rec. Bot. Survey India* **8**: 153–161.

Blatter, E. 1928. *Beautiful flowers of Kashmir*, vol. **1**. John Bale, sonsand Danielsson, London, W.I.

Boissier, E. 1872. *Flora Orientalis* **2**: 205–498. Geneva, H. Georg,

Borissova, A. G. 1946. *Astragalus* & *Acanthophace*. In: Komarov, V. *et al.*: *Flora U.R.S.S.*, vol. **12**, 918 p. Moskau, Leningrad.

Bunge, A. 1868. Generis Astragali species gerontogeae. *Mem. Acad. Imp. Sci. Saint Petersbourg* **11** (16): 1–140.

Bunge, A. 1869. Generis Astragali species gerontogeae. *Mem. Acad. Imp. Sci. Saint Petersboug* **15**: 1–254.

Chamberlain, D.F. and Matthews, V.A. 1969. Materials for a flora of Tuekey. XVIII. Leguminosae: *Astragalus. Notes Roy. Bot. Gard. Edinburgh* **29**: 285–310.

Chamberlain, D.F. & Mathews, V.A. 1970. *Astragalus*. In: Davis, P.H., *Flora of Turkey and the East Aegean Islands* **3**: 49–254. University Press, Edinburgh.

Chandra Sekar, K & Srivastava, S.K. 2009. *Flora of the Pin Valley National Park, Himachal Pradesh*. Botanical Survey of India, Kolkata.

Chater, A.O. 1968. *Astragalus*. In: Tutin T.G. *et al.* (eds.), *Flora Europaea* **2**: 108–124.

Chater, A. O., 1979. *Astragalus*. In: Williams L.H.J. and Hara, H. (eds.), An Enumeration of the Flowering Plants of Nepal, vol. **2**. British Museum (Natural History), London.

Chaudhary, L.B. 2005. Extended distribution of some *Astragalus* L. (Fabaceae) in India. *Rheedea* **15** (1): 75–79.

Chaudhary, L.B. 2006a. *Astragalus lachungensis* (Fabaceae), a new species from Sikkim Himalaya, India. *J. Jap. Bot.* **81** (3): 168 –172.

Chaudhary, L.B. 2006b. *Astragalus kongrensis* Benth. ex Baker (Fabaceae), A New Record for Central and North-West Himalayas. *Taiwania* **51** (3): 327–239.

Chaudhary, L.B. 2006c. *Astragalus zemuensis* W.W. Smith (Fabaceae) belongs to *Oxytropis* DC. *J. Jap. Bot.* **81** (5): 278–281.

Chaudhary, L.B. 2007. *Astragalus gooraiensis* (Fabaceae), A new species from Jammu and Kashmir in India. *Novon* **17**(4): 417–420.

Chaudhary, L.B, Anand, K.K. and Srivastava, R.K. 2007. Taxonomic study of endemic species of *Astragalus* L. (Fabaceae) of India. *Taiwania* **52** (3): 216–237.

Chaudhary, L.B., and Khan, Z.H. 2003. First report of *Astragalus* L. (Fabaceae) from Arunachal Pradesh, India. *Rheedea* **13** (1&2): 73–76.

Chaudhary, L.B. and Khan, Z.H. 2004. New distributional sites of some uncommon *Astragalus* L. (Fabaceae) in India. *Phytotaxonomy* **4**: 57–63.

Chaudhary, L.B. and Khan, Z.H. 2005a. *Astragalus khasianus* Benth. ex Bunge (Leguminosae), a new record to Myanmar. *J. Jap. Bot.* **80** (1): 52–56.

Chaudhary, L.B and Khan, Z.H. 2005b. *Astragalus* L. section *Chlorostachys* (Leguminosae -Papilionoideae) in India. In: Pandey, A.K., Wen, J. and Dogra, V.V (eds.), *Plant Taxonomy: Advances and Relevance*. CBS Publishers and Distributors, New Delhi, pp. 379–393.

Chaudhary, L.B. and Khan, Z.H. 2005c. Two new additions to the genus *Astragalus* L. (Leguminosae-Papilionoideae) of India. *Bull. Bot. Surv. India* **47** (1): 53–58.

Chaudhary, L.B. and Khan, Z.H. 2005d. A new species of *Astragalus* L. (Fabaceae) from Indo-Nepal region. *Rheedea* **15** (2): 129–131.

Chaudhary, L.B. and Khan, Z.H. 2006. *Astragalus uttaranchalensis* (Leguminosae-Papilionoideae), a new species from the Himalaya in India. *Taiwania* **51** (1): 36–40.

Chaudhary, L.B., Rana, T.S. and Anand. K.K. 2008. Current status of the systematics of *Astragalus* L. (Fabaceae) with special reference to the Himalayan species in India. *Taiwania* **53** (4): 338–355.

Chaudhary, L.B., Rana, T.S., Narzary, D. and Verma, S. 2007. A new species of *Astragalus* L. (Leguminosae) from India based on morphological and molecular markers. *Bot. J. Linn. Soc.* **154**: 27–34.

Chaudhary, L.B. and Rao, R.R. 2002. A Conspectus of the Genus *Astragalus* L. (Leguminosae- Papilionoideae) in India. In: Rao, R.R. & Chaudhary, L.B. (eds.), *Advances in Legume Research in India*. Bishen Singh Mahendra Pal Singh India. pp. 59–72.

Chaudhary, L.B. and Srivastava S.K. 2007. Taxonomic and Distributional notes on some *Astragalus* L. (Fabaceae) in India. *Taiwania* **52** (1): 25–48.

Chowdhery, H.J. 1984. Papilionaceae. In: Chowdhery, H.J. & Wadhwa, B.M., *Flora of Himachal Pradesh* **1**: 179– 234. Botanical Survey of India, Howrah.

Chowdhery, H. J. and Rao, R. R. 1990 (1992). Plant life in the Himalayan cold deserts: some adaptive strategies. *Bull. Bot. Surv. India* **32** (1–4): 43–56.

Chowdhery, H. J., Uniyal, B.P. and Balodi, B. 1992 (1997). A new species of *Astragalus* L. (Leguminosae) from Ladakh (J & K) India. *Bull. Bot. Surv. India* **34** (1–4): 209–211.

Collett, C.S.H. 1921. *Flora Simlensis*. Thacker, Spink & Co. London.

De Candolle, A. P. 1802. Astragalogia nempe *Astragali, Biserrulae et Oxytropidis*, nec non *Phacae, Colutae et Lessertiae* historia iconibus illustrate, Parisiis J.B. Garney, 269 p, 50 tab.

De Candolle, A.P. 1825. *Prodromus systematic naturalis regni vegetabilis*, vol. **2**. Treuttle & Wurtz, Paris.

Deml, I. 1972. Revision der Sektionen *Acanthophace* Bunge und *Aegacantha* Bunge der Gattung *Astragalus* L. *Boissiera* **21**: 1–235.

Dhar, U. and Kachroo, P. 1983. *Alpine Flora of Kashmir Himalaya*. Scientific Publishers, Jodhpur, India.

Dhillon, K.B. & Bajwa, P.S. 1969. A contribution to the botany of Ganganagar district, North Rajasthan. *Bull. Bot. Surv. India* **11**(3–4): 234–244.

Doyle, J.J., Doyle, J.L., Ballenger, J.A., Dickson E.E., Kajita, K. and Ohashi, H. 1997. A phylogeny of the chloroplast gene *rbc*L in the Leguminosae: taxonomic correlations and insights into the evolution of nodulation. *American J. Bot.* **84**: 541–554.

Engel, T. 1991. The evolution of rachis thorns in *Astragalus and Astracantha* (Leguminosae) and the systematic applicability of thorn anatomy. *Fl. Veg. Mundi* **9**: 17–27.

Fischer, F.E.L. 1853. Synopsis *Astragalorum Tragacantharum*. *Bull. Soc. Imp. Naturalistes Moscou* **26** (2): 316–486.

Frodin, D.G. 2004. History and concept of big plant genera. *Taxon* **53**: 753–776.

Gazer, M. 1993. Revision of *Astragalus* L. sect. *Sesamei* DC. (Leguminosae). *Sendtnera* **1**: 69–155.

Gentry, H.S. 1957. Gum Tragacanth in Iran. *Econ. Bot.* **11**: 40–63.

Ghahremani-Nejad, F. 2004. The sections of *Astragalus* L. with bifurcating hairs in Iran. *Turk. J. Bot.* **28:** 101–117.

Gillett, J. B. 1963. *Astragalus* L. (Leguminosae) in the highlands of tropical Africa. *Kew Bull.* **17** (3): 413–423.

Gomez-Sosa, E. 1979. Las especies Sudamericans del genero *Astragalus* (Leguminosae) I. Las especies Patagonicas Argentinas. *Darwiniana* **22**: 313–376.

Gomez-Sosa, E. 1981. Novedades en el genero *Astragalus* I (Leguminosae-Galegeae). *Darwiniana* **23**: 507–516.

Gomez-Sosa, E. 1982. Novedades en el genero *Astragalus* II (Leguminosae-Galegeae). *Darwiniana* **24**: 23–31.

Gray, A. 1864. A revision and arrangement (mainly by the fruit) of the North American species of *Astragalus* and *Oxytropis*. *Proc. Amer. Acad. Arts* **6**: 188–236.

Grierson, A.J.C and Long, D.C. 1987. Leguminosae. In: Grierson, , A.J.C and Long, D.C., *Flora of Bhutan*, vol. **1**, pt. **3**. R. B. G., Edinburgh.

Hu, J.-M., Sanderson, J.M., and Wojciechowski, M.F., 1999. Website for the largest genus of vascular plants – *Astragalus*. http://loco.biosci.arizona.edu/astragalus/astragalus_home.htm.

Johnston, I. M. 1947. *Astragalus* in Argentina, Bolivia and Chile. *J. Arnold Arboretum* **28**: 336–409.

Jones, M. E. 1923. *Revision of North American species of Astragalus*. Salt Lake City, Utah.

Kachroo, P., Sapru, B.L., and Dhar, U. 1977. *Flora of Ladakh*. International Book distributors, Dehra Dun, pp. 68–73.

Kang, Y. and Zhang, M.-L. 2004. Study of pollen brush in selected species of *Astragalus* L. subgenus *Pogonophace* Bunge (Leguminosae). *Plant Syst. Evol.* **249**: 1–8.

Kang, Y., Zhang, M.-L. and Chen, Z.-D. 2003. A preliminary phylogenetic study of the subgenus *Pogonophace* (*Astragalus*) in China based on ITS sequence data. *Acta Bot. Sin.* **45**: 140–145.

Karamian, R. and Ranjbar, M. 2005. *Astragalus* sect. *Asatragalus* (Fabaceae) in Iran. *Bot. Journ. Linn. Soc.* **147**: 363–368.

Kazemi, M., Kazempour Osaloo, S., Maassoumi, A.A and Rastegar Pouyani, E. 2009. Molecular phylogeny of selected Old World *Astragalus* (Fabaceae): incongruence among chloroplast trnL-F, ndhF and nuclear ribosomal DNA ITS sequences. *Nordic J. Bot.* **27:** 425–436.

Kazempour Osaloo, S., Maassoumi, A.A. and Murakani, N. 2003. Molecular systematics of the genus *Astragalus* L. (Fabaceae): phylogenetic analysis of nuclear ribosomal DNA internal transcribed spacers and chloroplast gene ndhF sequences. *Plant Syst. Evol.* **242** (1–4): 1–32.

Kazempour Osaloo, S., Maassoumi, A.A. and Murakani, N. 2005. Molecular systematics of the Old world *Astragalus* (Fabaceae) as inferred from nrDNA ITS sequence data. *Brittonia* **57** (4): 367–381.

Kitamura, S. 1964. *Flowering plants of west Pakistan and Afghanistan.* Results of the Kyoto University Scientific expedition to the Karakoram and Hindukush, 1955, vol. **3**. Kyoto.

Knaus, B.J., Cronn, R.C. and Liston A. 2005, Genetic characterization of three varieties of *Astragalus lentiginosus* (Fabaceae). *Brittonia* **57** (4): 334–344.

Kress, J., De Fillips, R.A., Farr, E. and Kyi, D.Y.Y. 2003. A checklist of Trees, Shrubs, Herbs and Climbers of Mayanmar. *Smith. Inst. Contr. United . States Nat. Herb.* **45**: 1–590.

Kumar, S. and Sane, P.V. 2003. *Legumes of South Asia: A Checklist.* Royal Botanic Gardens, Kew.

Kumar, V.S. and Karthikeyan, S. 2013. *Astragalus ramakianus*, a new name for the Persian *Astragalus turgidus* Podlech (Leguminosae). *Phytotaxa* **155** (1): 71–72.

Ledingham G. F. 1960. Chromosome numbers in *Astragalus* and *Oxytropis*. *Canadian J. Genet. Cytol.* **2**: 119–128.

Lewis, G.P., Schrire, B.D., Mackinder, B.A. and Lock, M. 2005. *Legumes of the world.* Royal Botanic Gardens, Kew.

Li, P.-C. and Ni, C.-C. 1979. New taxa of the genus *Astragalus* L. from Tibet. *Acta Phytotax. Sin.* **17**: 104–112.

Linnaeus, C. 1753. *Species Plantarum.* Stockholm.

Liston, A. 1989. Why are there so many species of *Astragalus* ? *Amer. J. Bot.***76** (supplement): 256.

Liston, A. 1990. Taxonomic notes on *Astragalus* section *Leptocarpi* subsection *Californici* (Fabaceae). *Brittonia* **42**(2): 100–104.

Liston, A. 1992. Variation in the chloroplast genes *rpo*C1 and *rpo*C2 of the genus *Astragalus* (Fabaceae): evidence from restriction site mapping of a PCR amplified fragment. *American J. Bot.* **79**: 953–961.

Liston, A. 1995. Use of the polymerase chain reaction to survey for the loss of the inverted repeat in the legume chloroplast genome. In: Crisp M. D. and Doyle J. J. (eds.), *Advances in legume systematics-phylogeny.* Royal Botanical Gardens, Kew, part **7**, pp. 31–40.

Liston, A. and Wheeler, J.A. 1994. The Phyllogenetic position of the genus *Astragalus* (Fabaceae): evidence from the chloroplast genes *rpo* C_1 and *rpo* C_2. *Biochem. syst. Ecol.* **22**: 377–388.

Lock, J.M. and Simpson, K. 1991. *Legumes of west Asia, a checklist*, Royal Botanical Gardens, Kew.

Lock, M. and Schrire, B.D. 2005. Galegeae. In: Lewis, G.P., Schrire, B.D., Mackinder, B.A. and Lock, M. (eds.), *Legumes of the world* . Royal Botanic Gardens, Kew, pp. 475–481.

Luo, M.-C., Hwu, K.-K. and Huang, T.-C. 2000. Taxonomic study of Taiwan *Astragalus* based on genetic variation. *Taxon* **49**: 35–40.

Maassoumi, A.A. 1990. Illustrated guide of the genus *Astragalus* in Iran. vol. **1**. Unpaged, 100 tables. Tehran. Research Institute of Forests and Rangelands Publ. no. **69**.

Maassoumi, A.A. 1993. Illustrated guide of the genus *Astragalus* in Iran. vol. 2. Unpaged, 100 tables. Tehran. Research Institute of Forests and Rangelands Publ. no. **86**.

Maassoumi, A.A. 1998. *Astragalus* L. *in the World, check list*, 618 p. Research Institute of Forests and Rangelands, Tehran.

Maassoumi, A.A. 2000. Important notes on *Astragalus* subgenus *Tragacantha* Bunge from Iran. *Iran. J. Bot.* **8**: 309–325.

Maassoumi, A.A. 2003. Papilionaceae (*Astragalus* I). *Flora of Iran*, vol. **43**, 386 p. Tehran.

Maassoumi, A.A. and Podlech, D. 1990. New species of *Astragalus* L. (Leguminosae) from Iran. *Mitt. Bot. Staatss. Munchen* **29**: 509–514.

Mabberley, D.J. 1997. The Plant-Book, a portable dictionary of the vascular palnts, ed. 2, Cambridge University Press, Cambridge, UK.

Mehrnia, M., Zarre, S. and Sokhan-Sanj, A. 2005. Intra- and inter-specific relationship within the *Astragalus microcephalus* complex (Fabaceae) using RAPD. *Biochem. Syst. Ecol.* **33**: 149–158.

Naithani, B.D. 1984 – 1985. *Flora of Chamoli*. **1**: 143–147. B.S.I., Howrah.

Ohashi, H. 1975. Leguminosae In: Ohashi, H., *Flora of Eastern Himalaya*, University of Tokyo Press, Japan.

Ohba, H., Akiyama, S. and Wu, S.-K. 1994. *Astragalus* (Fabaceae): Taxonomic Studies of the plants from the Kunlun Mountains in Central Asia, 1. *J. Jpn. Bot.* **70** (1): 11–31.

Ott. E. 1978. Revision der Sektion *Chronopus* Bge. Der Gattung *Astragalus* L. *Phanerogam. Monog.* **9**: 1–142. Vaduz.

Pallas, P. 1800–1802. *Species Astragalorum*. G. Martini, Lipsiae.

Parker, R.N. 1921. N.W. Himalayan *Astragali* of the subgenus *Aegacantha*. *Kew Bull. Misc. Inform.* **1921**: 260–270.

Parker, R.N. 1924. *A forest flora of Punjab with Hazara and Delhi*, ed. 2, 138–143.

Parker, R.N. 1924. A new species of *Astragalus* from Kumaon. *Indian Forester*. **49**: 78–79.

Pirani, A., Zarre, S., Tillich, H.J., Podlech, D. and Niknam, V. 2006. Spine anatomy and its systematic application in *Astragalus* sect. *Rhacophorus* s. l. (Fabaceae) in Iran. *Flora* **201**: 240–247.

Podlech, D. 1967. Neue und bemerkenswerte Fabaceae aus Nordost-Afghanistan (Beitrage zur Flora von Afghanistan II). *Mitt. Bot. Staatss. Munchen* **6**: 547–591.

Podlech, D. 1973. Neue und bemerkenswerte *Asragalus*-Arten aus Afghanistan (Beitrage zur Flora von Afghanistan IV). *Mitt. Bot. Staatss. Munchen* **11**: 259–321.

Podlech. D. 1975 a. Revision der Sektion *Stipitella* G. Grig. ex Podlech der Gattung *Astragalus. Mitt. Bot. Staatss. Munchen* **12**: 33–50.

Podlech, D. 1975 b. Revision der Sektion *Caraganella* Beg. der Gattung *Astragalus* L. *Mitt. Bot. Staatss. Munchen* **12**: 153–166.

Podlech, D. 1982. Neue aspekte zur evolution und gliederung der gattung *Astragalus* L. *Mitt. Bot. Staatss. Munchen* **18**: 359–378.

Podlech, D. 1983. Zur taxonomie und nomenclatur de tragacanthoiden Astragali. *Mitt. Bot. Staatss. Munchen* **19**: 1–23.

Podlech, D. 1984. Revision von *Astragalus* L. sect. *Herpocaulos* Bunge. *Mitt. Bot. Staatss. Munchen* **20**: 441–449.

Podlech, D. 1986, Taxonomic and phytogeographical problems in *Astragalus* of old world and south-west Asia. *Proc. Roy. Soc. Edinburgh* **89** B: 37– 43.

Podlech, D. 1988. Revision von *Astragalus* L. sect. *Caprini* DC. (Leguminosae). *Mitt. Bot. Staatss. Munchen* **25**: 1– 924.

Podlech, D. 1990. Revision von *Astragalus* L. sect. *Platyglottis* Bunge (Leguminosae). *Mitt. Bot. Staatss. Munchen* **29**: 541–572.

Podlech, D. 1991. The Systematics of the annual species of the genus *Astragalus* L. (Leguminosae). *Fl. Veg. Mundi* **9**: 1– 8.

Podlech, D. 1993a. Revision von *Astragalus* L. Sektion *Eremophysa* Bunge und *Eremophysopsis* Gontsch. (Leguminosae). *Sendtnera* **1**: 45–64.

Podlech, D. 1993b. Beittrage zur K enntnis der gattung *Astragalus* L. (Leguminosae) IV. Neue namen und Kombinationen in den Gattungen *Astragalus* L. und *Astracantha* Podlech: V. Uber einige bisher nomenklatorisch nicht berucksichtigte *Astragalus*- Namen von Linne. *Sendtnera* **1**: 269–271.

Podlech, D. 1994. Revision der altweltlichen annuellen Arten der Gattung *Astragalus* L. (Leguminosae). *Sendtnera* **2**: 39–170.

Podlech, D. 1998a, Phylogeny and progression of characters in Old World Astragali (Leguminosae). In: Zhang, A. and Wu, S. (eds.), *Floristic characteristics and diversity of east Asian palnts.* China Higher Education Press, Beijing, pp. 405– 407.

Podlech, D. 1998b. Typification of *Astragalus* species II. Species mainly of the Herbaria of Paris (P) and Geneva (G). *Sendtnera* **5**: 247–263.

Podlech, D. 1999a. New Astragali and *Oxytropis* from north Africa and Asia, including some new combinations and remarks on some species. *Sendtnera* **6**: 135–191.

Podlech, D. 1999b. Papilionaceae III: *Astragalus* I. In: Rechinger, K. H. (ed.), *Flora Iranica*. Akademische Druck-u. Verlagsanstalt, Graz, No. **174**, pp. 1–350, tabs. 227.

Podlech, D. 2001. Contribution to the knowledge of the genus *Astragalus* L. (Leguminosae) VII–X. *Sendtnera* **7**: 163–201.

Podlech, D. 2008. *Thesaurus Astragalorum*. Index of all taxa within the genus *Astragalus* L. and other genera but belonging to the genus *Astragalus*. Taxa of the Old World and related taxa of the New World. http://www.botanik.biologie.uni-muenchen.de/botsyst/theslit.html.

Podlech, D. 2008. The genus *Astragalus* (Fabaceae) in Europe with exclusion of the former Soviet Union. *Feddes Repert.* **119** (5-6): 310–387.

Podlech, D. 2009. Some new species and combinations in *Astragalus* and Phyllolobium (Fabaceae). *Feddes Repert.* **120** (1-2): 48–58.

Podlech, D. 2010. Four new *Astragalus* species (legiminosae) from China and Bhutan. *Pak. J. Bot.*, Special Issue (S.I. Ali Festschrift) **42:** 27–34.

Podlech, D. and Deml, I. 1967. Eine interessante neue Astragalus-Art aus Afghanistan. (Beitrage zur Flora van Afghanistan) *Mitt. Bot. Staassammi. Munchen* **6**: 541–546.

Podlech, D. and Ekici, M. 2008. Some new and interesting *Astragalus* species (Fabaceae) from Turkey. *Feddes Repert.* **119** (1-2): 24–36.

Podlech, D. and Maassoumi, A.A. 2003. New species of *Astragalus* L. (Fabaceae) from Iran, mainly of sects. *Incani* and *Malacothrix. Feddes Repert.* **114**: 320–349.

Podlech, D., Maassoumi, A.A. and Zarre, Sh. 2012. *Flora Iranica* **179:** Papilionaceae VII, *Astragalus* V. 312 p., 37 tab., Akademische Verlagsgesellschaft Salzburg.

Podlech, D. and Sytin, A. 1996: Typification of Russian and some other species of *Astragalus* I. *Sendtnera* **3**: 149–176.

Podlech, D. and Sytin, A. 2002. New species of *Astragalus* L. (Leguminosae) sect. *Hololeuce, Onobrychoidei, Ornithopodium* and *Synochreati* and a new section *Baldaccia*. *Sendtnera* **8**: 155–166.

Podlech, D. and Xu, L.-R. 2004. New species and combinations in *Astragalus* (Leguminosae) from China and the Himalayas. *Novon* **14**: 216–226.

Podlech, D. and Xu, L.-R. 2007. New species and new combination in *Astragalus* (Leguminosae) from China. *Novon* **17**: 228–254.

Podlech, D. and Zarre, Sh. (with collaboration of Ekici, M., Maassoumi, A.A. and Sytin, A.) 2013. *A taxonomic revision of the genus Astragalus L. (Leguminosae) in the Old World*, vols. 1–3. Naturhistorisches Museum Wien-Austria.

Podlech, D. , Zarre, Sh. and Maassoumi, A.A. 2001. *Flora Iranica* **175**: Papilionaceae IV, *Astragalus* II. 197 p., 135 tab., Akademische Verlagsgesellschaft Salzburg.

Podlech, D. , Zarre, Sh., Maassoumi, A.A., Ekici, M. and Systin, A. 2010. *Flora Iranica* **178:** Papilionaceae VI, *Astragalus* IV. 805 p., 375 tab., Akademische Verlagsgesellschaft Salzburg.

Polhill, R.M. 1981. Tribe *Galegeae*. In: Polhill, R.M. and Raven, P.H. (eds), *Advances in Legume Systematics* **1**: 357–363. Royal Botanic Gardens, Kew, England.

Press, J.R., Shrestha, K.K. and Sutton, D.A. 2000. *Annotated checklist of the flowering plants of the Nepal*. The Natural History Museum, London.

Qian, Z.-G. and Yang, Y.W. 2003. New materials of *Astragalus* (Leguminosae) from Yunnan. *Acta Bot. Yunnanica* **25**: 283–284.

Ranjbar, M. and Karamian, R. 2002. *Astragalus* sect. *Astragalus* (Fabaceae) in Iran, complementary notes with a key to the species. *Nordic J. Botany* **22**: 177–181.

Ranjbar, M. and Maassoumi, A.A. 1998 (1997). New species and new records of the genus *Astragalus* (Leguminosae) from Iran. *Iran. J. Bot.* **7**: 235–238.

Ranjbar, M., Maassoumi, A. A. and Podlech, D. 2002. *Astragalus* sect. *Alopecuroidei* (Fabaceae) in Iran, complementry notes with a key to the species. *Willdenowia* **32**: 85–91.

Rao, R.R. and Balodi, B. 1989. Two new species of *Astragalus* L. from Western Himalaya. *Proc. Nat. Acad. Sci., India* **59** (B) IV: 475–478.

Rechinger, K.H. 1958 (1957). *Leguminosae*. In: Koeie, M. & Rechinger, K.H., Symbolae Afghanicae III. *Biol. Skr. Dan. Vid. Selsk.* **9** (3): 1–208.

Rechinger, K.H. 1964. Papilionaceae. In: Rachinger, K.H., *Flora of Lowland Iraq*. Hafner publishing co., New York. pp. 334–389.

Reer, U. & Podlech, D. 1986: Die europäischen Vertreter der Gattung *Astracantha* Podl. (*Leguminosae*). *Mitt. Bot. Staatss. München* **22**: 513–569.

Riahi, M., Zarre, S., Chehregani, A. and Shahsavan-Behboudi, B. 2003. Seed development in two species of medifixed hairy *Astragalus* (Fabaceae). *Flora* **198:** 211–219.

Riahi, M., Zarre, S., Maassoumi, A. A., Kazempour Osaloo, S. and Wojciechowski, M.F. 2011. Towards a phylogeny for *Astragalus* section *Caprini* (Fabaceae) and its allies based on nuclear and plastid DNA sequence. *Plant Syst. Evol.* **293:** 119–133.

Rydberg, P.A. 1929. Astragalanae. *North American Flora* **24**: 251– 462.

Rydberg, P.A. 1929. Genera of North American Fabaceae VI. *Astragalus* and related genera (continued). *Amer. J. Bot.* **16** (4): 197–206.

Rydberg, P.A. 1930. Genera of North American Fabaceae VII. *Astragalus* and related genera (continued). *Amer. J. Bot.* **17** (3): 231–238.

Sanderson, M.J. 1991. Phyllogenetic relationships within North American *Astragalus* L. (Fabaceae) *Syst. Bot.* **16**: 414–430.

Sanderson, M.J. and Doyle, J.J. 1993. Phylogenetic relationships in North American *Astragalus* L. (Fabaceae) based on choloroplast DNA restriction site variation. *Syst. Bot.* **18**: 395–408.

Sanderson, M.J. and Liston, A. 1995. Molecular phyllogenetic systematics of *Galegeae*, with special reference to *Astragalus*. In: Crisp, M.D. and Doyle, J.J. (eds.), *Advances in legume Systematics- phylogeny*. Royal Botanic Garden, Kew, part **7**, 331–350.

Sanderson, M.J. and Wojciechowski, M.F. 1996. Diversification rates in a temperate legume clade: Are there "So many species" of *Astragalus* (Fabaceae)?. *Amer. J. Bot.* **83** (11): 1488–1502.

Sanjappa, M. 1992. *Legumes of India*, Bishen Singh Mahendra Pal Singh, Dehra Dun.

Scherson, R.A., Choi, H.-K., Cook, D.R. and Sanderson M.J. 2005. Phylogenetics of new world *Astragalus*: Screening of novel nuclear loci for the reconstruction of phylogenies at low taxonomic level. *Brittonia* **57** (4): 354–366.

Scherson, R.A., Vidal, R. and Sanderson M.J. 2008. Phylogeny, biogeography and rates of diversification of New World *Astragalus* (Leguminosae) with an emphasis on South American radiations. *Amer. J. Bot.* **95:** 1030–1039.

Sharma, M.P., Aswal, B.S. and Mehrotra, B.N. 1990. *Astragalus badrinathensis* (Fabaceae): A new species from Chamoli district, Uttar Pradesh, India. *J. Econ. Tax. Bot.* **14** (1): 113–114.

Simpson, N.D. 1915. An Enumeration of the Chinese *Astragali*. With Descriptions of New Species. *Notes Roy. Bot. Gard. Edinburgh* **8**: 240–264.

Sirjaev, G. 1939. Conspectus Tragacantharum (*Astragalus* L. subgenus *Tragacantha* Bge). *Feddes Report. Spec. Nov. Regni Veg.* **47**: 194–208, 225–261.

Sirjaev, G. 1951. New Species of Fabaceae. *Brittonia* **7**: 275–277.

Sirjaev, G. and Rechinger, K.H. 1951. *Astragali et Onobrychides* novi orientales. *Ann. Naturhist. Mus. Wien* **58**: 2–16.

Smith, W.W. and Cave, G.H. 1911. The vegetation of the Zemu and Llonakh valley of Sikkim. *Rec. Bot. Surv. India* **4** (5): 185–186.

Spellenberg, R. 1976. Chromosome-numbers and their cytotaxonomic significance for North-American *Astragalus* (Fabaceae). *Taxon* **25**: 463–476.

Srivastava, T.N., Badola, D.P., and Gupta, O.P. 1984. New records of flowering plants for Ladakh-II. *Indian J. Forestry* **7**: 335–336.

Sytin, A.K. 1991: Generis *Astragalus* L. (Fabaceae) sectionis *Myobroma* (Stev.) Bunge specierum caucasicorum revisio. *Novosti Sist. Vyssh. Rast.* **28**: 102–110.

Sytin, A.K. and Laktionov, A.P. 2007. Notes on *Astragalus* (Fabaceae) from the Astrakhan region. *Bot. Zhurn.* **92**(6): 905–912.

Taeb, F., Zarre, S., Podlech, D., Tillich, H.-J., Kazempour Osaloo, S. and Maassoumi, A.A. 2007. A contribution to the phylogeny of annual species of *Astragalus* (Fabaceae) in the Old World using hair micromorphology and other morphological characters. *Feddes Repert.* 118 (5-6): 206–227.

Taubert, P. 1891. Leguminosae. In: Engler & Prantl, *Die naturilichen Pflanzenfamilien* **3**(3): 70 –185. Leipzig.

Tietz, S. 1988: Revision von *Astragalus* L. sect. *Campylanthus* Bunge, sect. *Microphysa* Bunge und sect. *Poterion* Bunge. *Mitt. Bot. Staatss. München* **27**: 135–380.

Tietz, S. 1996: Bemerkungen zu *Astragalus* L. sect. *Tricholobus* Bunge (*Fabaceae*). *Sendtnera* **3**: 177–186.

Tietz, S. and Zarre, Sh. 1994. Revision von *Astragalus* L. sect. *Megalocystis* Bunge (Fabaceae). *Sendtnera* **2**: 287–363.

Torrey, J. and Gray, A. 1838. Tribe *Astragaleae*. *A Flora of North America* **1**: 328–353.

Townsend, C. C. 1974. Leguminosae. In: Townsend, C.C. (ed.), *Flora of Iraq* **3**: 231–442.

Ulbrich, E. 1921. Leguminosae asiaticae novae vel criticae. *Notizbl. Bot. Gart. Berlin- Dahlem* **8**: 83–89.

Ulziykhutag, N. 2001. Novaja sektzija I novye vidy podroda *Cercidothrix* Bunge roda *Astragalus* (Fabaceae) iz tzentralnoj asii. [Sectio nova et species novae subgeneris *Cercidothrix* Bunge generis *Astragalus* L. (Fabaceae) ex Asia centrali]. *Novosti Sist. Vyssh. Rast.* **33**: 135–145.

Valsechi, F. 1994. About the *Astragalus tragacantha* complex (Leguminosae) in the Mediterranean. *Webbia* **49** (1): 31–41.

Verbeken, D., Dierckx, S. and Dewettinck, K. 1838. Exudate gums: occurrence, production and application. *App. Microbiol. Biotechnil.* **63**: 10–21.

Wenninger, J. 1991. Revision von *Astragalus* L. sect. *Chlorostachys* Bunge, sect. *Phyllolobium* Bunge und sect. *Skythropos* Simpson (Leguminosae). *Mitt. Bot. Staatss. München* **30**: 1–196.

Wojciechowski, M.F. 2005. *Astragalus* (Fabaceae): A molecular phylogenetic perspective. *Brittonia* **57** (4): 382–396.

Wojciechowski, M.F., Sanderson, M.J., Baldwin, B.G. and Donoghue, M.J. 1993. Monophyly of aneuploid *Astragalus* (Fabaceae): Evidence from nuclear ribosomal DNA internal transcribed spacer sequences. *Amer. J. Bot.* **80**: 711–722.

Wojciechowski, M.F., Sanderson, M.J. and Hu, J.-M. 1999. Evidence on the monophyly of *Astragalus* (Fabaceae) and its major subgroups based on nuclear ribosomal DNA ITS and chloroplast DNA *trn*L intron data. *Syst. Bot.* **24**: 409–437.

Wojciechowski, M.F., Sanderson, M.J., Steele, K.P. and Liston, A. 2000. Molecular phylogeny of the "temperate herbaceous tribes" of papilionoid legumes: a supertree approach. In: Herendeen, P.S. and Bruneau, A. (eds.), *Advances in legume systematic*, part **9**: 277–298. Royal Botanic Gardens, Kew.

Wu, Yu-Hu 1998. New taxa of the genus *Astragalus* from Qinghai. *Acta Bot. Yunnanica* **20**: 35–40.

Xu, L. R. and Podlech, D., 2006. Some new taxa of *Astragalus* L. (Leguminosae) from China and Nepal. *Feddes Repert.* **117** (3–4): 225–231.

Xu, L. and Podlech, D., 2010. *Astragalus*. In: Wu, Z.Y., Raven, P.H. and Hony, D.Y. (eds.), *Flora of China* (Fabaceae)**10**: 328–453. Science Press, Beijing and Mussouri Botanical Garden Press, St. Louis.

Zarre, Sh. 2000. Systematic revision of *Astragalus* sect. *Adiaspastus*, sect. *Macrophyllium* and sect. *Pterophorus*. *Englera* **18**: 1–219.

Zarre, Sh. 2003. Hair micromorphology and its phyllogenetic application in thorny species of *Astragalus* (Fabaceae). *Bot. J. Linn. Soc.* **143**: 323–330.

Zarre, Sh. and Azani, N. 2013. Perspective in taxonomy and phylogeny of the genus *Astragalus* (Fabaceae): a review. *Progr. Biolog. Sci.* **3**: 1–6.

Zarre, Sh., Maassoumi, A.A. and Podlech, D. 2008. *Flora Iranica* **177:** Papilionaceae V, *Astragalus* III. 124p., 45 tab. Verlag des Naturhist. Museums Wien.

Zarre, Sh. and Podlech, D. 1996. Taxonomic Revision of *Astragalus* L. sect. *Hymenostegis* Bunge (Leguminosae). *Sendtnera* **3:** 255–312.

Zarre, Sh. and Podlech, D. 1997. Problems in the taxonomy of tragacanthic Astragali. *Sendtnera* **4:** 243–250.

Zarre, Sh. and Podlech, D. 2001a. A short contribution to genus *Astragalus* L. (Fabaceae) in Turkey. *Pakistan J. Bot.* **33** (2): 153–155.

Zarre, Sh. and Podlech, D. 2001b. Taxonomic Revision of *Astragalus* sect. *Acanthophace* (Fabaceae). *Sendtnera* **7**: 233–255.

Zarre, Sh. and Podlech, D. 2002. *Astragalus* sect. *Semnanenses* (Fabaceae): a new monotypic section from Iran. *Nordic J. Botany* **21** (5): 485–491.

Zarre, Sh., Podlech, D. and Taeb, F. 2005. New species of the genus *Astragalus* L. (Fabaceae), mainly from the "Flora Iranica" area. *Feddes Repert.* **116**: 54–79.

Zhang, M.L. 2002. Systematics of *Astragalus* subgenus *Pogonophace* (Legumonisae). *Acta Bot. Yunnanica* **24** (5): 543–553.

Zhang, M.L. 2003. Biogeography of *Astragalus* subgenus *Pogonophace* (Legumonisae). *Acta Bot. Yunnanica* **25** (1): 25–32.

Zhang, M., Kang, Y., Zhou, L. and Podlech, D. 2009. Phylogenetic origin of *Phyllolobium* with a further implication for diversification of *Astragalus* in China. *J. Int. Plant. Bio.* **51:** 889–899.

Zhang, M.L. and Podlech, D. 2006. Revision of the genus *Phyllolobium* Fisch. (Leguminosae-Papilionoideae). *Feddes Repert.* **117** (1–2): 41–64.

Index to Botanical Names

Numbers refer to page numbers. Accepted names have been printed in italics-bold and synonyms in italics only. Page numbers in bold type denote main treatment.

Plates

Plate 1. ***Astragalus multiceps*** Wall. ex Benth.: **A**-Habit, **B**-Close-up of habit.

Plate 2. ***Astragalus oplites*** Benth. ex Parker: **A**-Habit (in flowers), **B**-Habit (in fruits).

Plate 3. ***Astragalus webbianus*** Grah. ex Benth.: **A**-Habit, **B**-Close-up of habit.

Plate 4. ***Astragalus zanskarensis*** Benth. ex Bunge: **A**-Habit, **B**-Close-up of habit.

Plate 5. ***Astragalus melanostachys*** Benth. ex Bunge: **A**-Habit, **B**-Close-up of habit.

Plate 6. ***Astragalus candolleanus*** Royle ex Benth.: **A**-Habit, **B**-Close-up of habit.

Plate 7. *Astragalus malacophyllus* Benth. ex Bunge: **A**-Habit, **B**-Close-up of habit.

Plate 8. ***Astragalus rhizanthus*** Royle ex Benth.: **A**-Habit (in flowers), **B**-Habit (in fruits).

Plate 9. ***Astragalus tecti-mundi*** Freyn subsp. ***orientalis*** Podlech: **A**-Habit, **B**-Close-up of habit.

Plate 10. *Astragalus coluteocarpus* Boiss. subsp. ***coluteocarpus***: **A**-Habit, **B**-Close-up of habit.

Plate 11. ***Astragalus anfractuosus*** Bunge: **A**-Habit, **B**-Close-up of flowers.

Plate 12. ***Astragalus chlorostachys*** Lindl.: **A**-Habit, **B**-Close-up of habit.

Plate 13. ***Astragalus concretus*** Benth.: **A**- Habit, **B**-Close-up of fruits.

Plate 14. ***Astragalus graveolens*** Buch.-Ham. ex Benth.: **A**-Habit, **B**-Close-up of habit (in fruits, photo by M. Sanjappa).

Plate 15. ***Astragalus stipulatus*** D. Don: **A**-Habit, **B**-Close-up of habit.

Plate 16. ***Astragalus uttaranchalensis*** L.B.Chaudhary & Z.H. Khan: **A**-Habit, **B**-Close-up of fruits.

Plate 17. ***Astragalus tibetanus*** Benth. ex Bunge: **A**-Habit, **B**-Close-up of habit.

Plate 18. ***Astragalus nivalis*** Kar. & Kir.: **A**-Habit, **B**-Close-up of habit.

Plate 19. ***Astragalus amherstianus*** Royle ex Benth.: **A**-Habit, **B**-Close-up of fruits.

Plate 20. ***Astragalus himachalensis*** L. B. Chaudhary & T. S. Rana: **A**-Habit, **B**-Close-up of habit.

Plate 21. ***Astragalus himalayanus*** Klotzsch: **A**-Habit, **B**-Close-up of habit.

Plate 22. ***Astragalus munroi*** Benth. ex Bunge.: **A**-Habit, **B**-Close-up of Habit, **C**-Close-up of fruits.

Plate 23. ***Astragalus leucocephalus*** Grah. ex Benth.: **A**-Habit, **B**-Close-up of habit.

Plate 24. ***Astragalus sikkimensis*** Benth. ex Bunge: **A**-Habit, **B**-Close-up of habit.

Plate 25. ***Astragalus strictus*** Grah. ex Benth.: **A**-Habit, **B**-Close-up of habit.

Plate 26. ***Astragalus tenuicaulis*** Benth. ex Bunge: Habit.

Plate 27. ***Astragalus falconeri*** Bunge var. ***falconeri***: **A**-Habit, **B**-Close-up of habit.

Plate 28. ***Astragalus hoffmeisteri*** (Klotzsch) Ali: **A**-Habit, **B**-Close-up of habit.

Plate 29. ***Astragalus verus*** Olivier: **A**-Habit, **B**-Close-up of habit.

Plate 30. ***Astragalus peduncularis*** Royle ex Benth.: **A**-Habit, **B**-Close-up of habit (in flowers), **C**-Close-up of habit (in fruits).

Plate 31. ***Phyllolobium donianum*** (DC.) M. L. Zhang & Podlech: **A**-Close-up of habit (in flowers), **B**-Close-up of habit (in fruits, photo by A. S. Kanthraj).

Plate 32. ***Astragalus grahamianus*** Royle ex Benth. (*Rao* 9367-BSD).

Plate 33. *Astragalus langtangensis* Podlech (*Schilling et al.* 380-KATH).

Plate 34. *Astragalus multiceps* Wall. ex Benth. (*Chaudhary et al.* 263803-LWG).

Plate 35. *Astragalus oplites* Benth. ex Parker (*Aswal & Goel* 16281-CDRI).

Plate 36. *Astragalus psilocentros* Fisch. (*Thomson s. n.*- K).

Plate 37. *Astragalus webbianus* Grah. ex Benth. (*Aswal* 10979-CDRI).

Plate 38. *Astragalus zanskarensis* Benth. ex Bunge (*Chaudhary et al.* 263844-LWG).

Plate 39. *Astragalus commixtus* Bunge (*Bhattacharyya* 47355-BSD).

Plate 40. ***Astragalus melanostachys*** Benth. ex Bunge (*Chaudhary et al.* 206749-LWG).

Plate 41. *Astragalus thomsonii* Podlech (*Bhattacharyya* 52199-BSD).

Plate 42. ***Astragalus candolleanus*** Royle ex Benth.(*Chaudhary & Khan* 206759-LWG).

Plate 43. *Astragalus drasianus* H. J. Chowdhery (*Nair* 22537-BSD).

Plate 44. *Astragalus malacophyllus* Benth. ex Bunge (*Khan* 223082-LWG).

Plate 45. *Astragalus pindreensis* Benth. ex Baker (*Falconer s.n.*-K).

Plate 46. ***Astragalus lessertoides*** Benth. ex Bunge (*Duthie* 984-BM).

Plate 47. *Astragalus pseudochlorostachys* Ali (*Schmis* 2166-BM).

Plate 48. ***Astragalus tecti-mundi*** Freyn subsp. ***orientalis*** Podlech (*Chaudhary et al. s.n.*-LWG).

Plate 49. *Astragalus coluteocarpus* Boiss. subsp. ***chitralensis*** Wenninger (*Stainton* 7806 -BM).

Plate 50. ***Astragalus coluteocarpus*** Boiss. subsp. ***chitralensis*** Wenninger (*Stainton* 7974-BM).

Plate 51. *Astragalus anfractuosus* Bunge (*Bhattacharyya* 49092-BSD).

Plate 52. *Astragalus chlorostachys* Lindl. (*Chaudhary et al.* 263807-LWG).

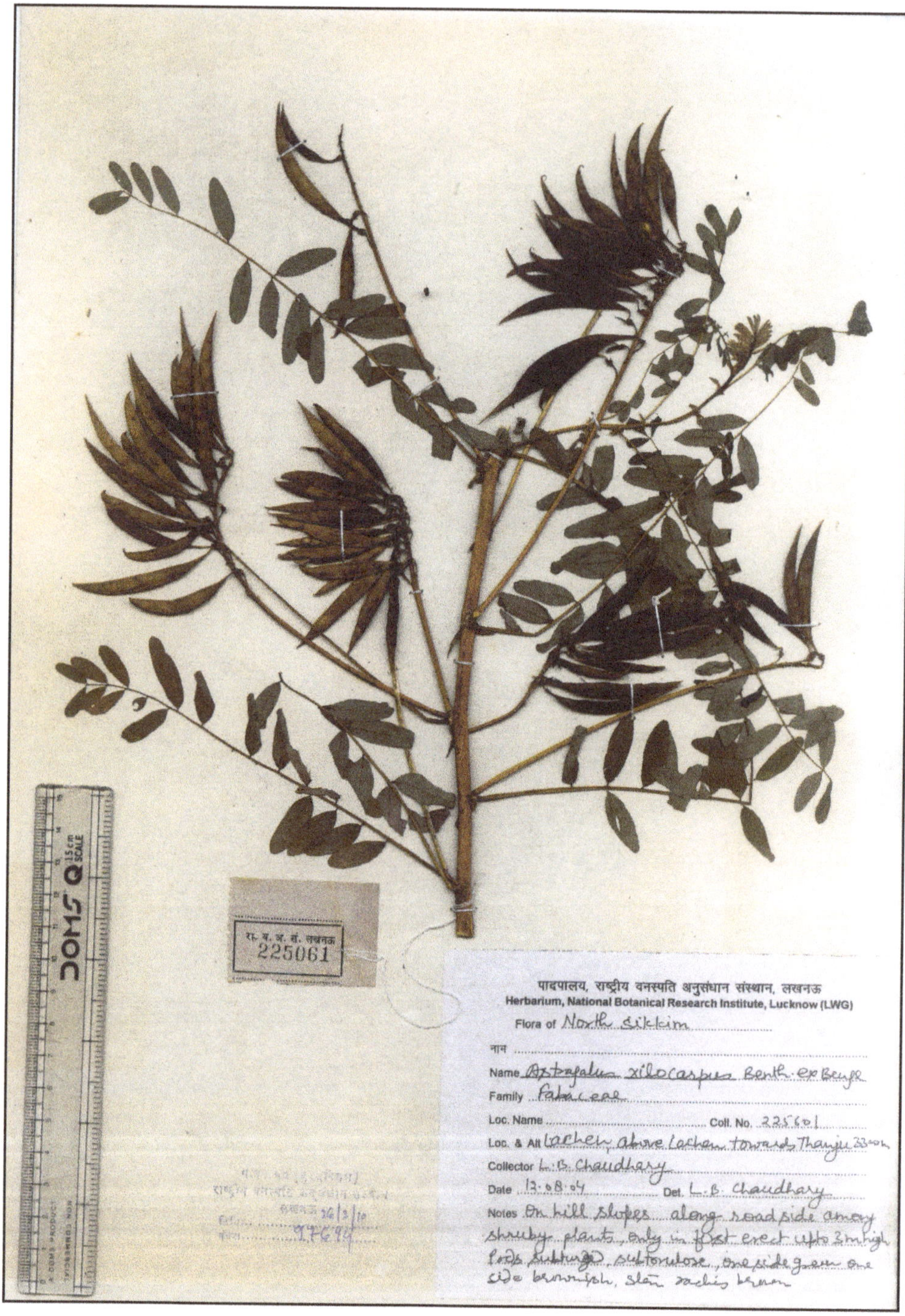

Plate 53. *Astragalus concretus* Benth. (*Chaudhary et al.* 225061-LWG).

Plate 54. ***Astragalus graveolens*** Buch.-Ham. ex Benth. (*Timins* 71-BM).

Plate 55. *Astragalus graveolens* Buch.-Ham. ex Benth.(*Prescott-Decie s. n.*- BM).

Plate 56. *Astragalus maxwelli* Royle ex Benth.(*Bhattacharyya* 71260-BSD).

Plate 57. *Astragalus stewartii* Baker (*Ludlow & Sherriff* 7726- BM).

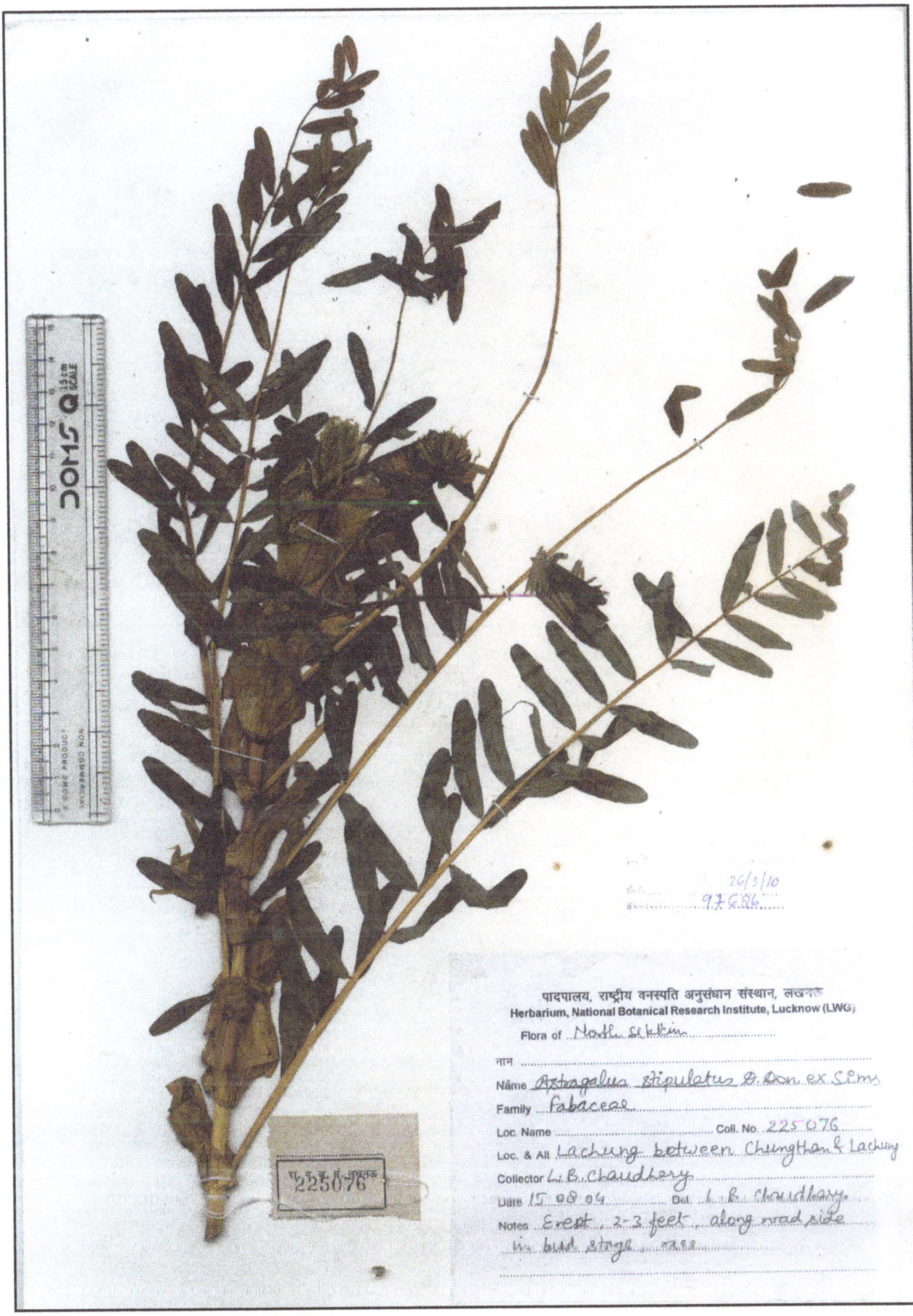

Plate 58. ***Astragalus stipulatus*** D. Don (*Chaudhary* 225076-LWG).

Plate 59. ***Astragalus uttaranchalensis*** L.B.Chaudhary & Z.H.Khan (*Khan* 223097-LWG).

Plate 60. *Astragalus scorpiurus* Bunge (*Ludlow & Sherriff* 8104-BM).

Plate 61. *Astragalus tibetanus* Benth. ex Bunge(*Bhattacharyya* 52237-BSD).

Plate 62. *Astragalus nivalis* Kar. & Kir.(*Chaudhary et al.* 206758-LWG).

Plate 63. *Astragalus amherstianus* Royle ex Benth. (*Aswal* 10468-CDRI).

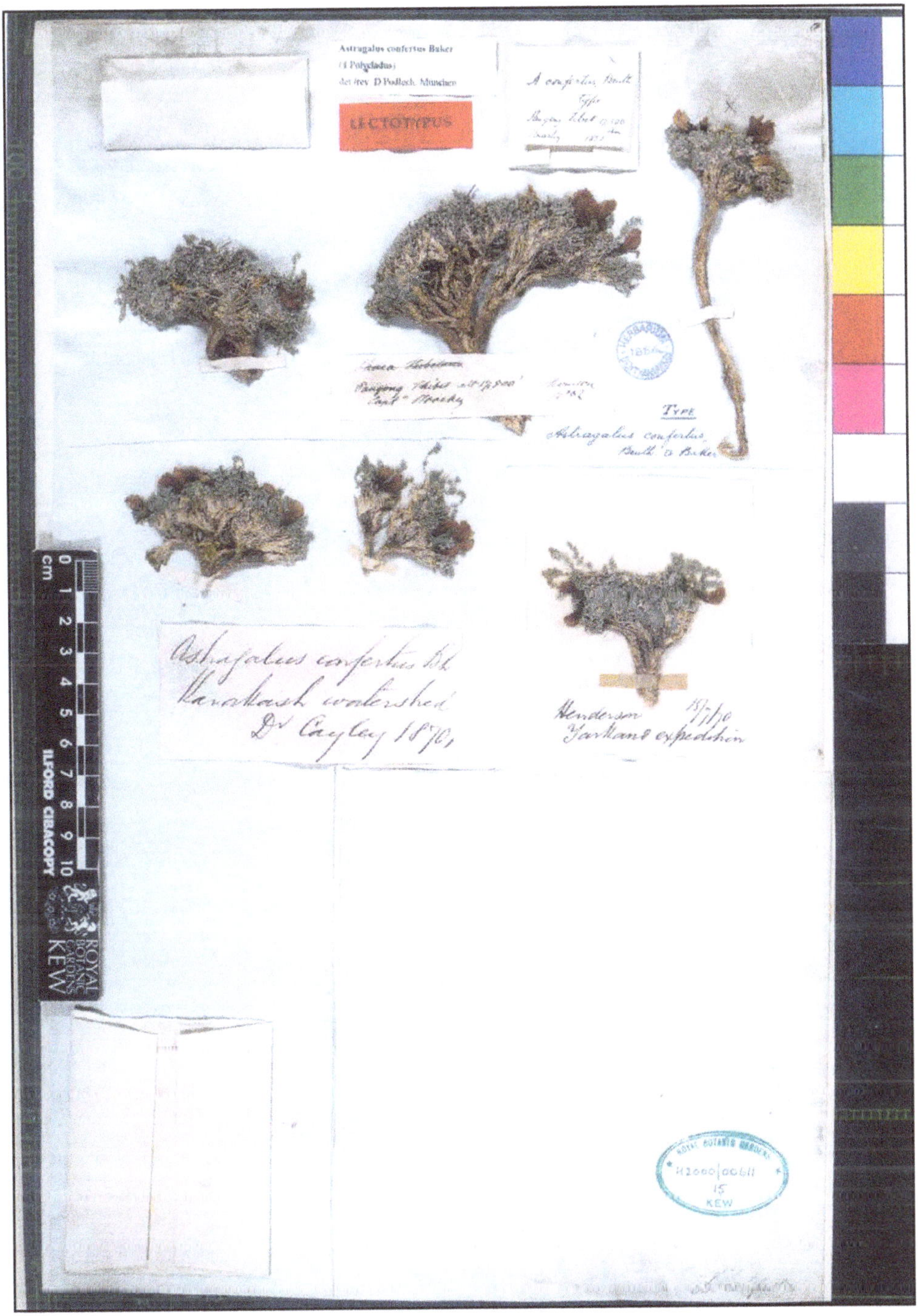

Plate 64. ***Astragalus confertus*** Benth. ex Bunge (*Strachey* 15-K).

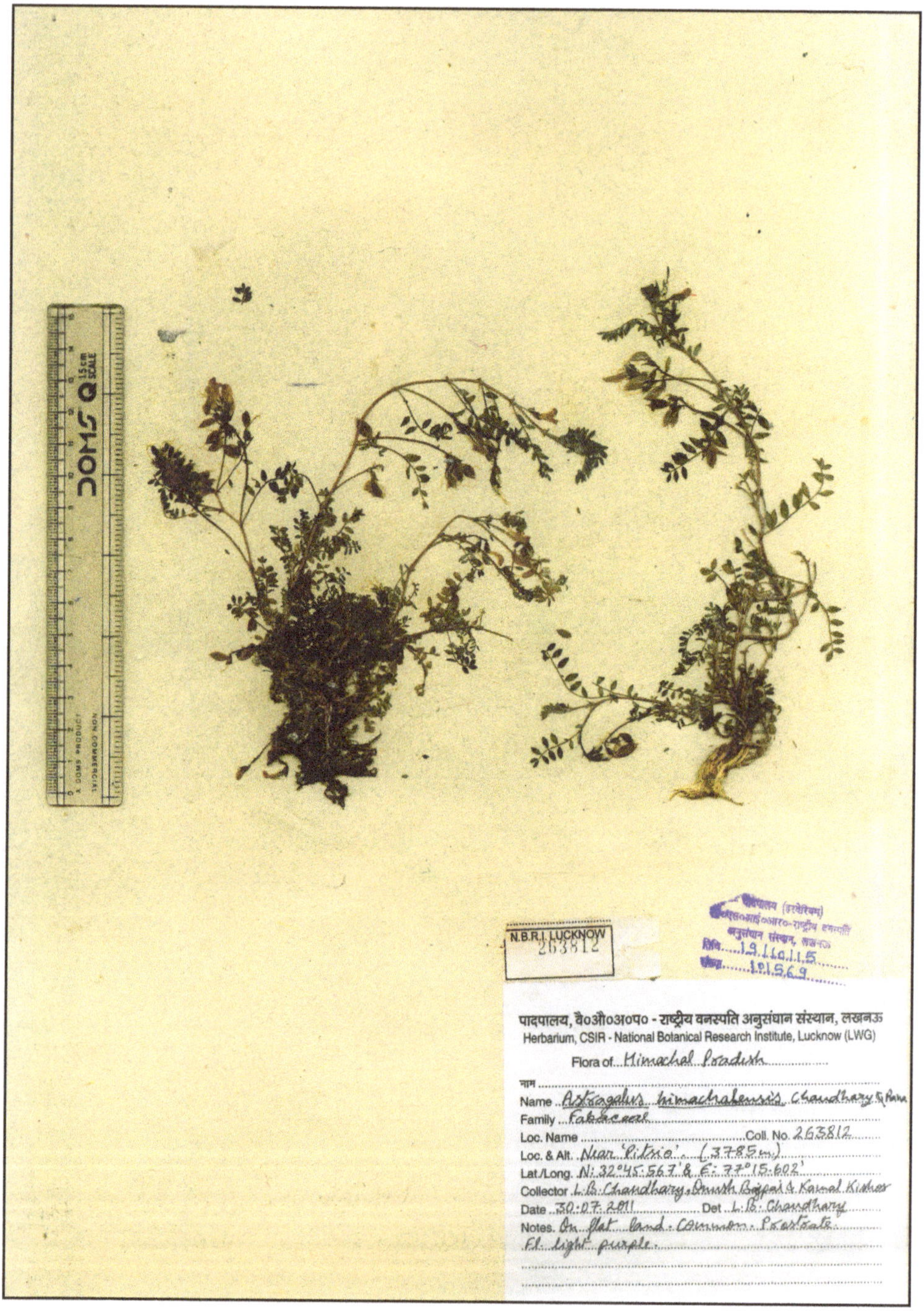

Plate 65. *Astragalus himachalensis* L.B. Chaudhary & T.S. Rana (*Chaudhary et. al.* 263812-LWG).

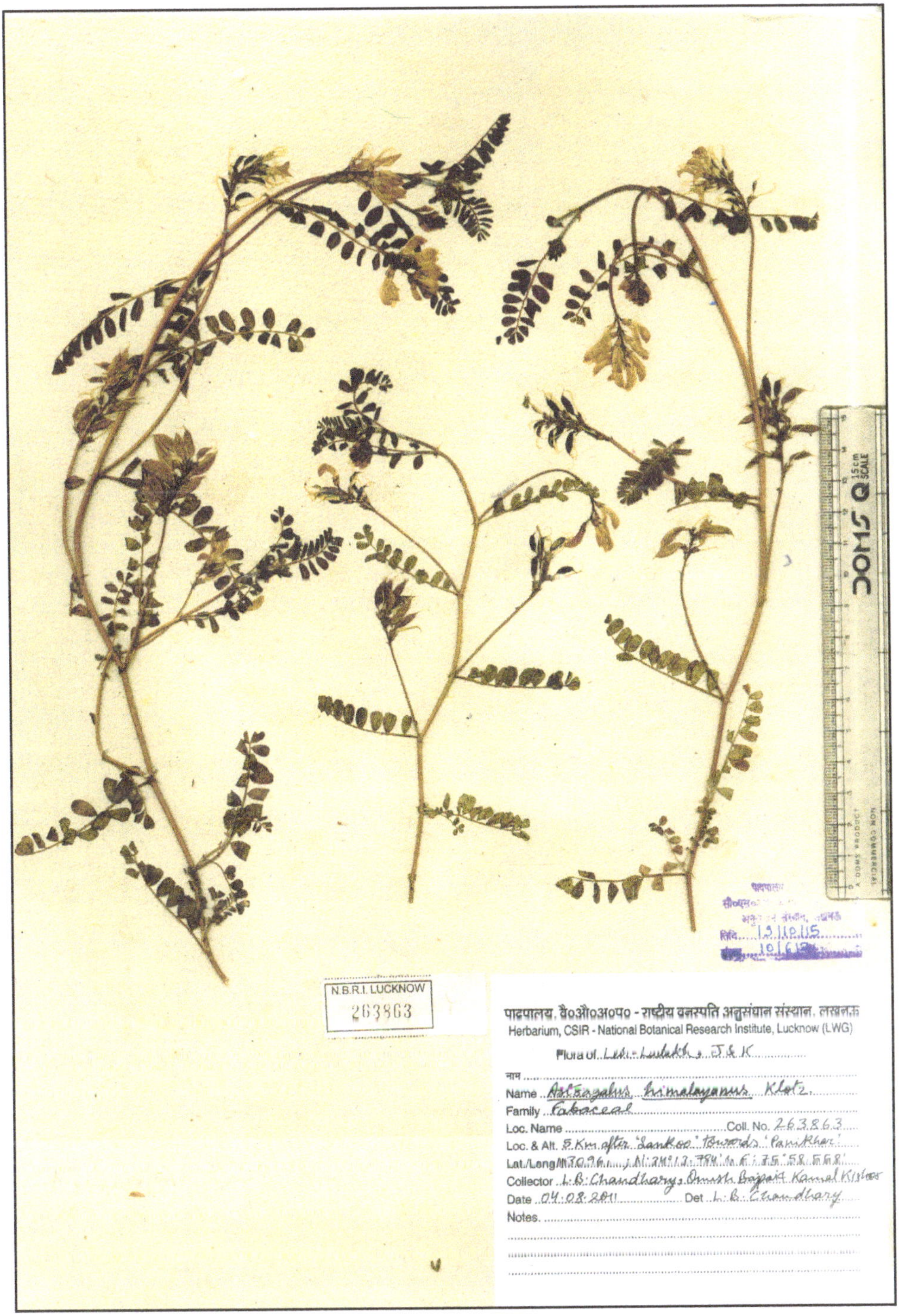

Plate 66. ***Astragalus himalayanus*** Klotzsch (*Chaudhary et al.* 263863-LWG).

Plate 67. *Astragalus kashmirensis* Bunge (*Falconer s. n.*-K).

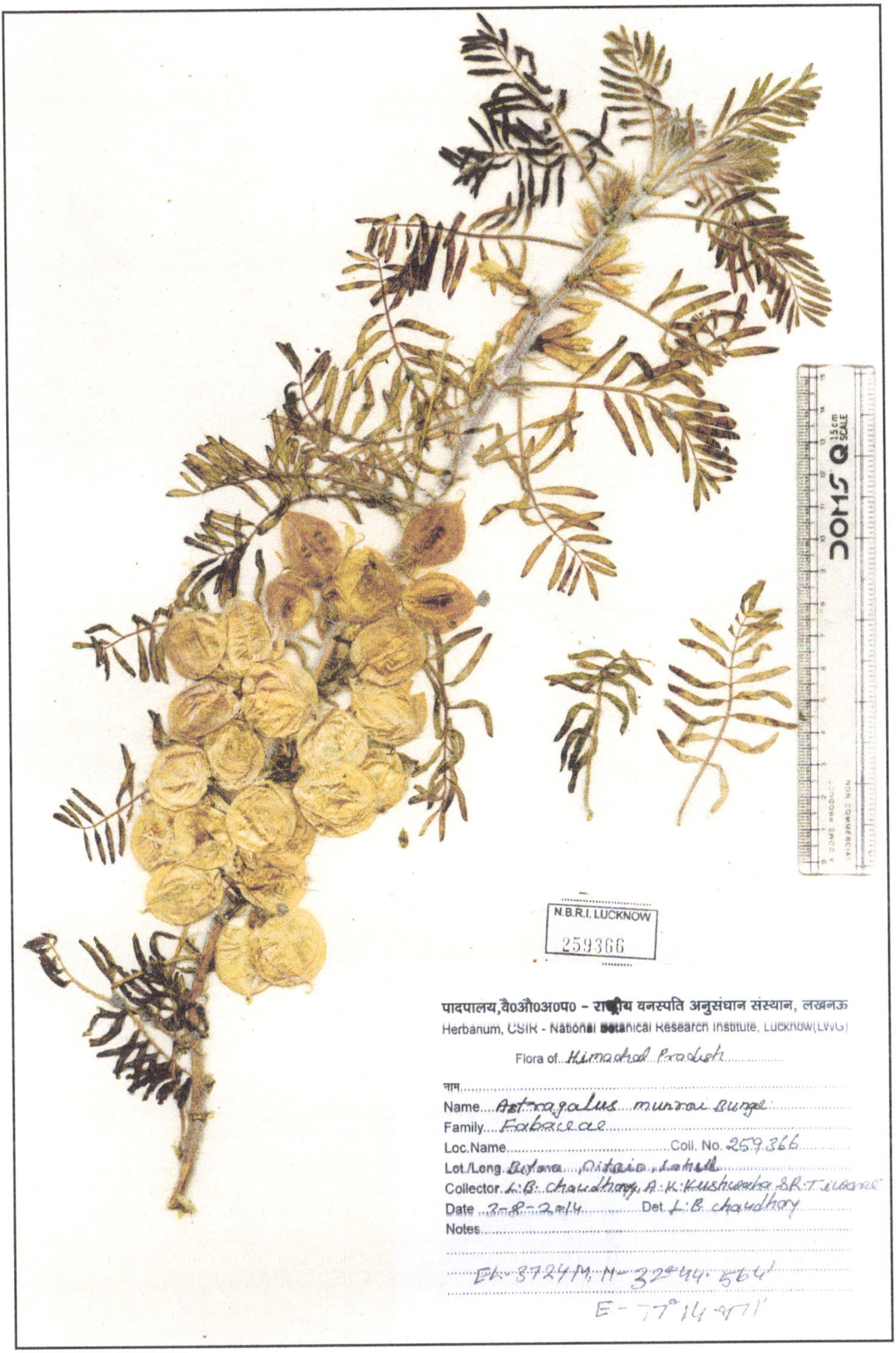

Plate 68. *Astragalus munroi* Benth. ex Bunge (*Chaudhary et al.* 259366-LWG).

Plate 69. *Astragalus pseudo-beckii* Sirjaev & Rech. f (*Stainton* 7912-BM).

Plate 70. *Astragalus ophiocarpus* Bunge (*Nair* 36527 BSD).

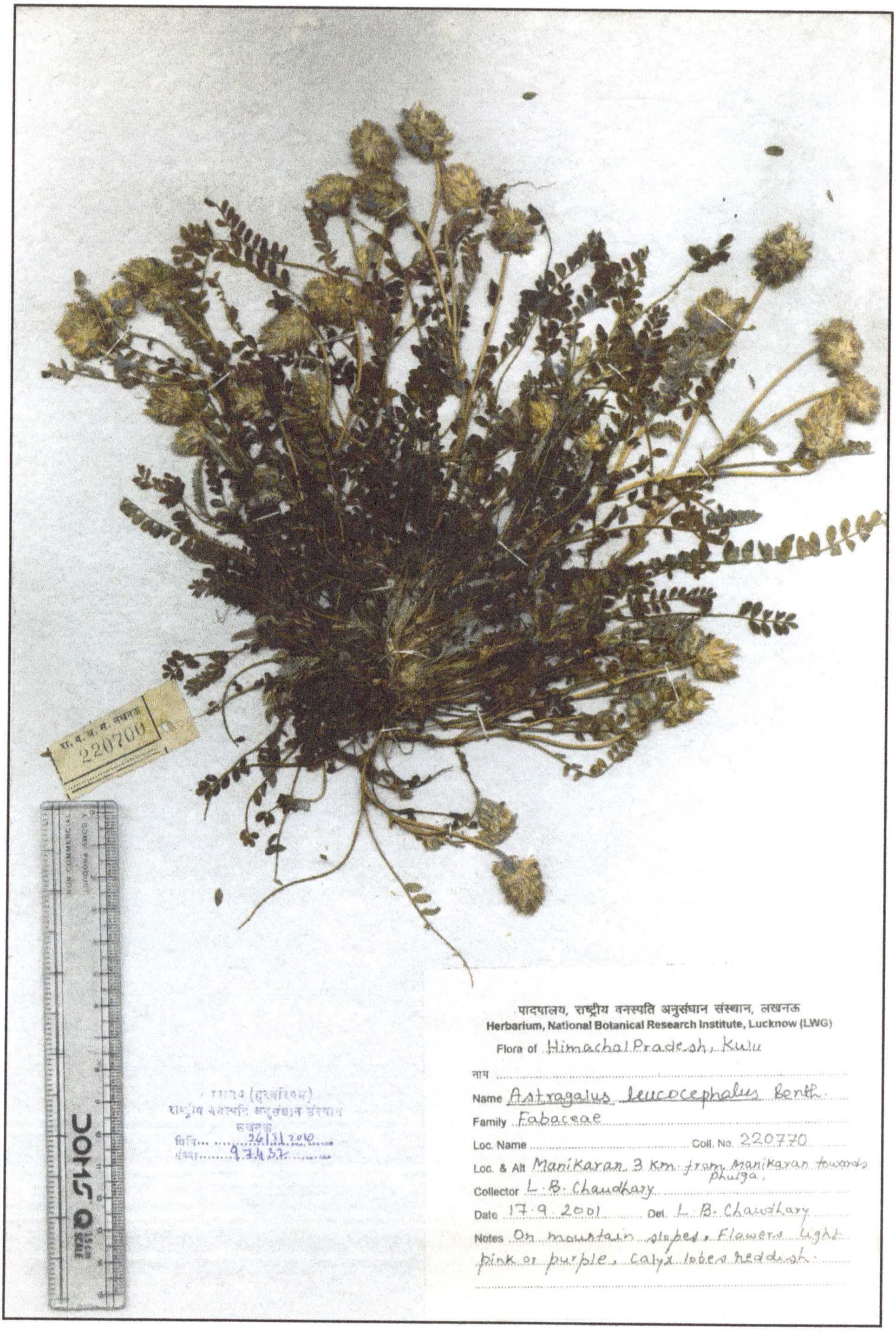

Plate 71. ***Astragalus leucocephalus*** Grah. ex Benth. (*Chaudhary* 220770-LWG).

Plate 72. ***Astragalus sanjappae*** L.B.Chaudhary & Z.H.Khan (*Chaudhary* 225071-LWG).

Plate 73. *Astragalus sikkimensis* Benth. ex Bunge (*Chaudhary* 225082-LWG).

Plate 74. *Astragalus strictus* Grah. ex Benth. (*Bhattacharyya* 49110-BSD).

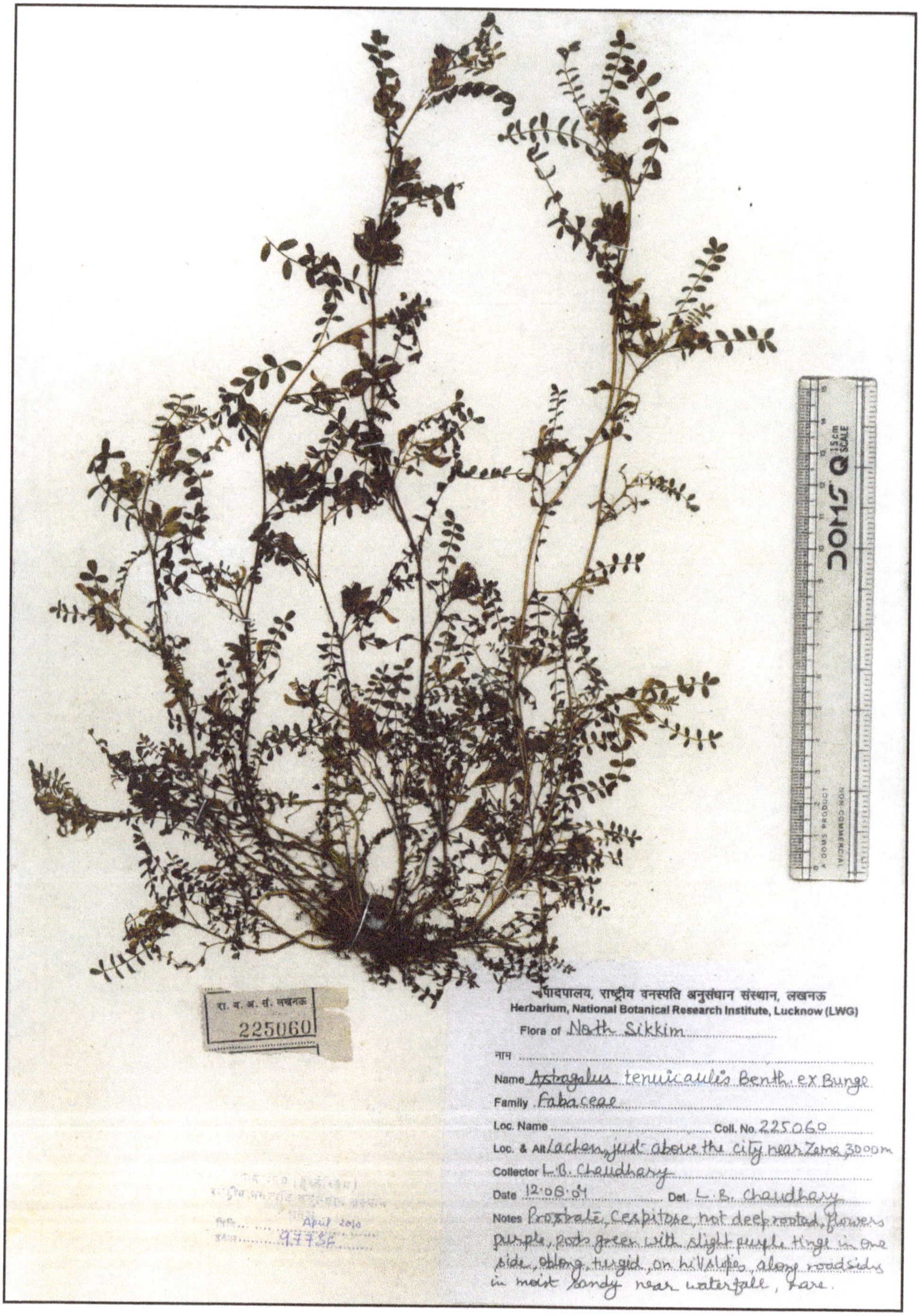

Plate 75. *Astragalus tenuicaulis* Benth. ex Bunge (*Chaudhary* 225060-LWG).

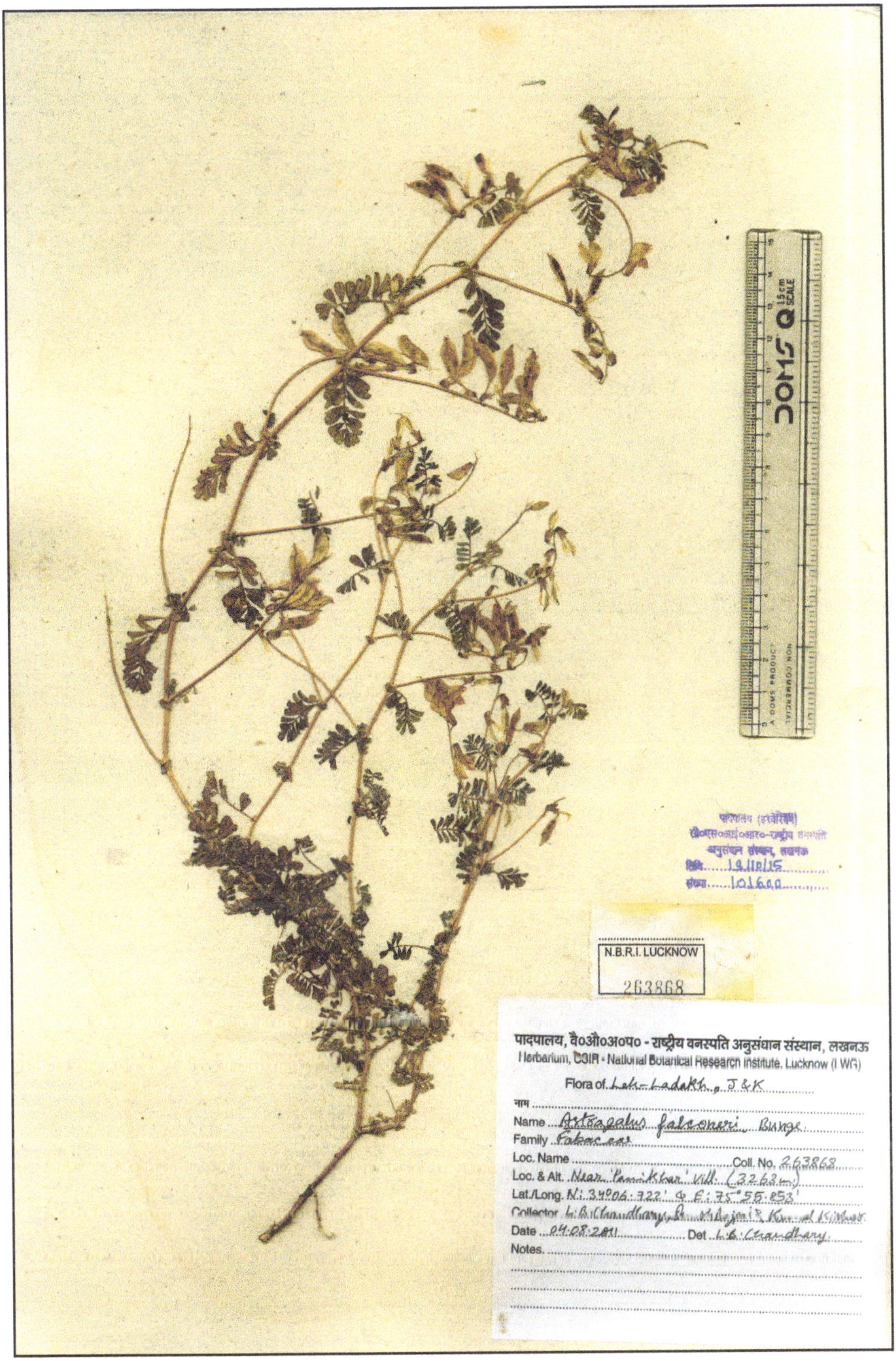

Plate 76. *Astragalus falconeri* Bunge var. ***falconeri*** (*Chaudhary et al.* 263868-LWG).

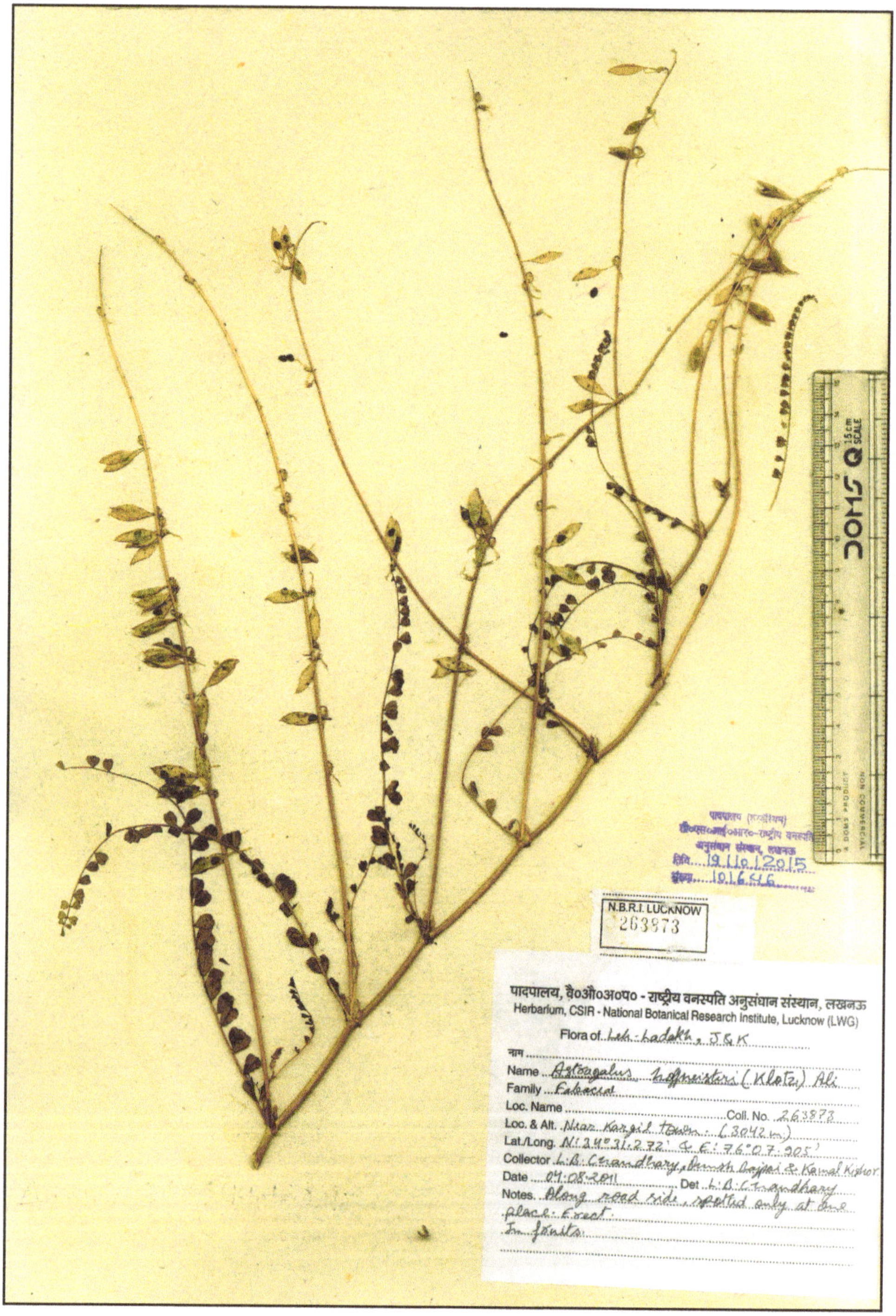

Plate 77. *Astragalus hoffmeisteri* (Klotzsch) Ali (*Chaudhary et al.* 263873-LWG).

Plate 78. ***Astragalus verus*** Olivier (*Schmid* 2383-BM).

Plate 79. *Astragalus tribuloides* Delile (*Thomson s.n.*-BM).

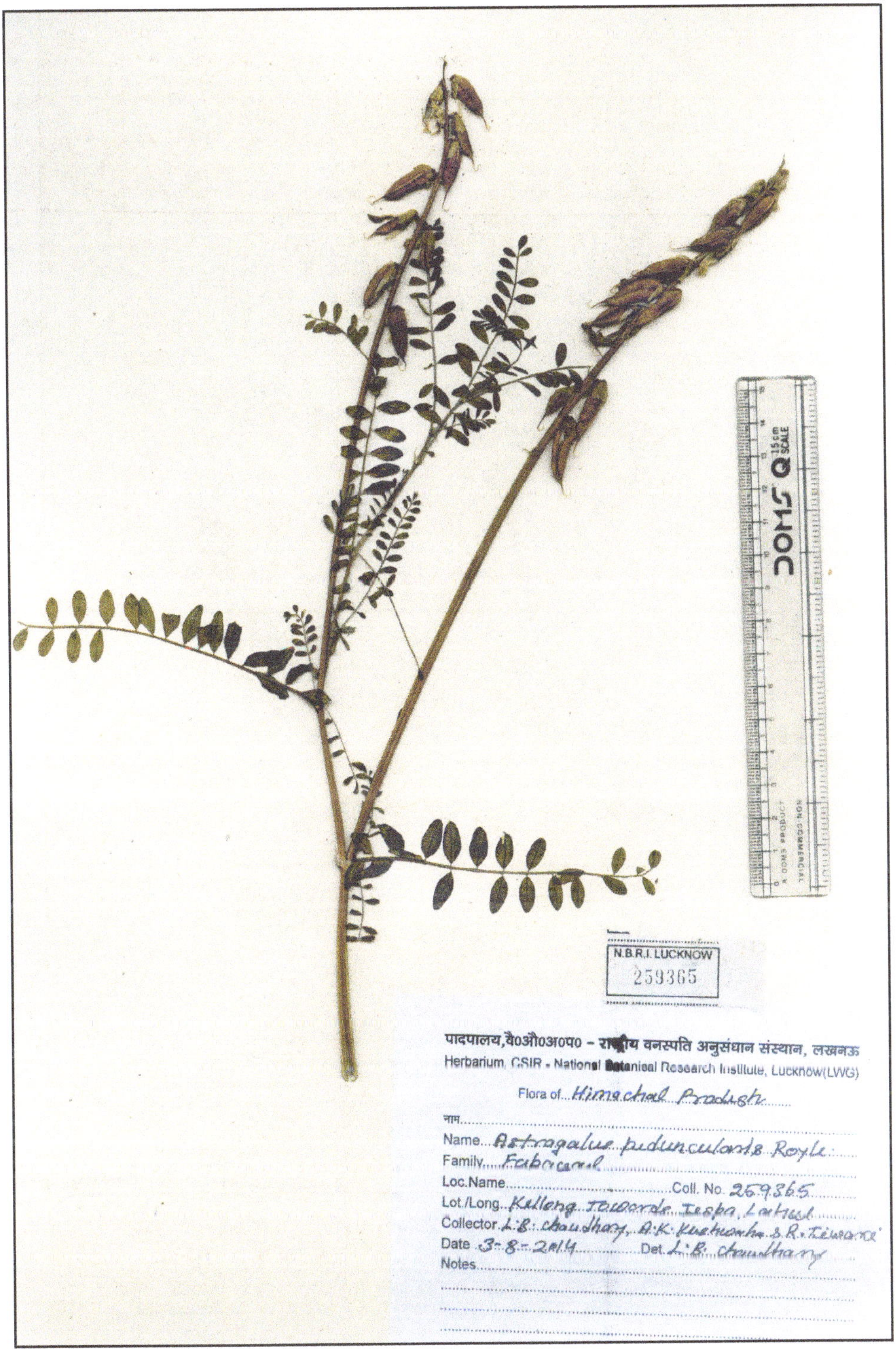

Plate 80. ***Astragalus peduncularis*** Royle ex Benth. (*Chaudhary et al.* 259365-LWG).

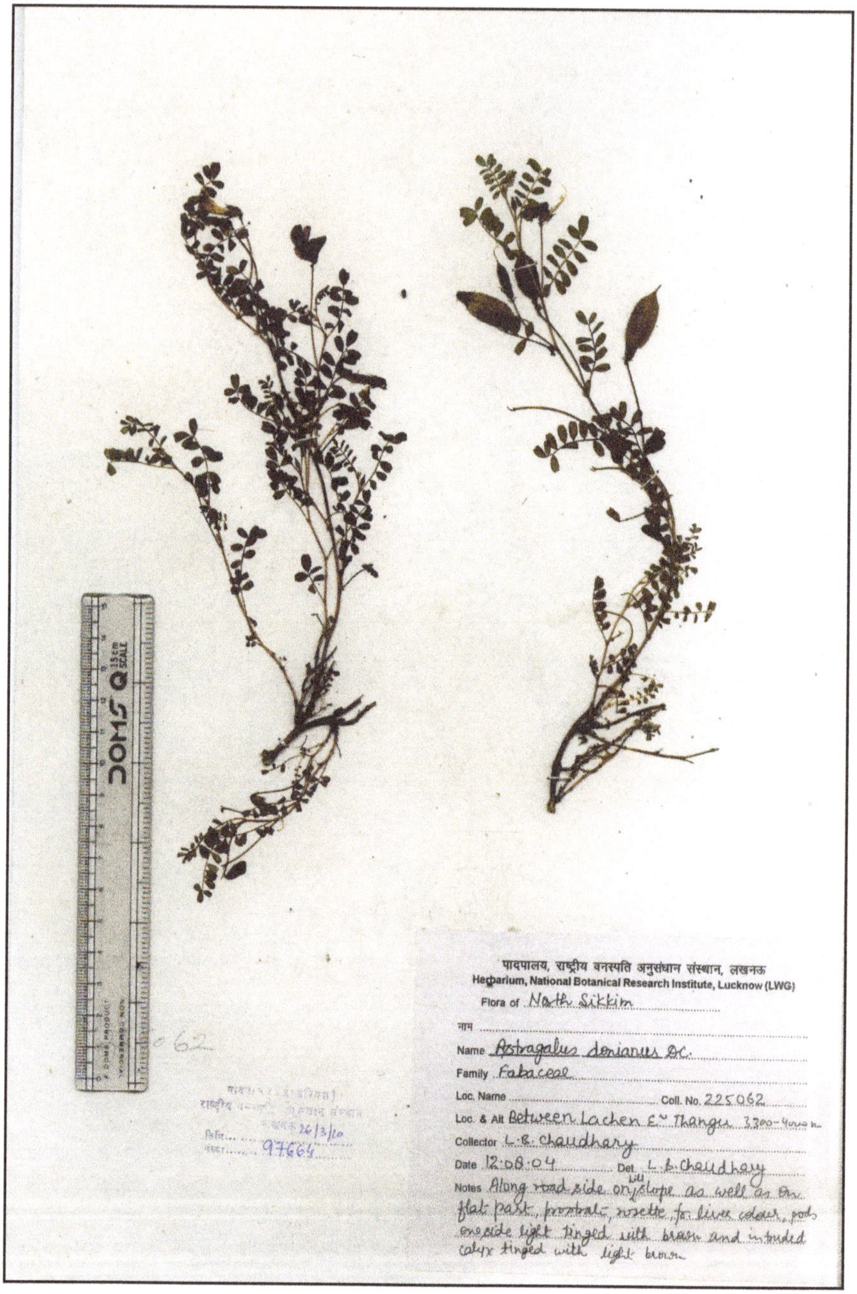

Plate 81. ***Phyllolobium donianum*** (DC.) M. L. Zhang & Podlech *(Chaudhary* 225062-LWG).

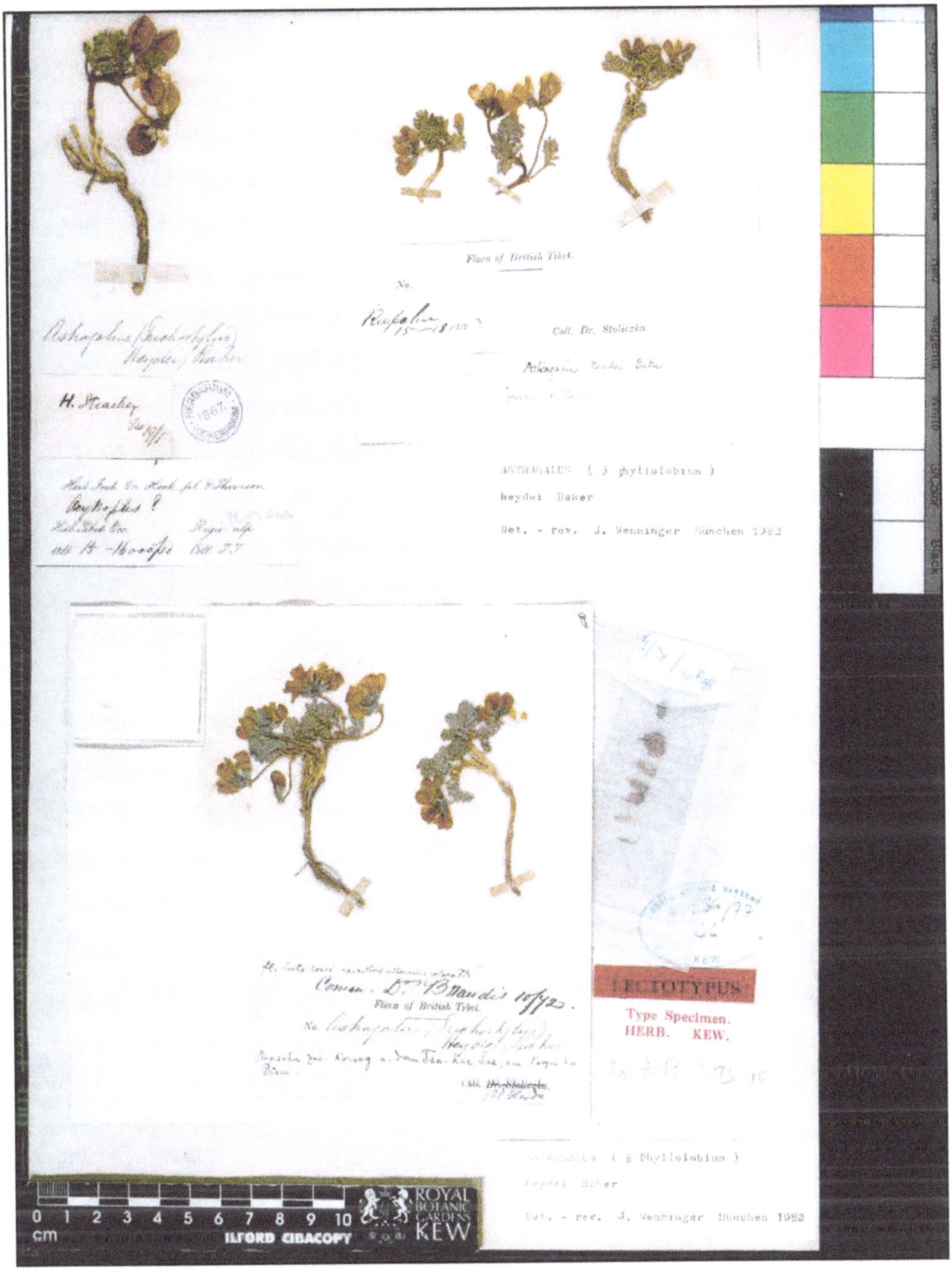

Plate 82. *Phyllolobium heydei* (Baker) M. L. Zhang & Podlech (*Heyde s. n.*-K).

Plate 83. *Phyllolobium tribulifolium* (Bunge) M. L. Zhang & Podlech (*Thomson s. n.*-K).